Lecture Notes in Computer Science 9964

Commenced Publication in 1973
Founding and Former Series Editors:
Gerhard Goos, Juris Hartmanis, and Jan van Leeuwen

Marko van Eekelen · Ugo Dal Lago (Eds.)

Foundational and Practical Aspects of Resource Analysis

4th International Workshop, FOPARA 2015
London, UK, April 11, 2015
Revised Selected Papers

Springer

Editors
Marko van Eekelen
Computer Science Department
Open University
Heerlen
The Netherlands

and

Digital Security
Radboud University
Nijmegen
The Netherlands

Ugo Dal Lago
Dipartimento di Informatica
Università di Bologna
Bologna
Italy

ISSN 0302-9743　　　　　　　ISSN 1611-3349　(electronic)
Lecture Notes in Computer Science
ISBN 978-3-319-46558-6　　　ISBN 978-3-319-46559-3　(eBook)
DOI 10.1007/978-3-319-46559-3

Library of Congress Control Number: 2016952506

LNCS Sublibrary: SL2 – Programming and Software Engineering

Printed on acid-free paper

This Springer imprint is published by Springer Nature
The registered company is Springer International Publishing AG Switzerland

Preface

This LNCS volume constitutes the formal proceedings of the 4th Foundational and Practical Aspects of Resource Analysis (FOPARA 2015) Workshop, which was held with the Developments in Implicit Computational Complexity (DICE 2015) Workshop at the European Joint Conferences on Theory and Practice of Software (ETAPS 2015) in London, UK, at the Queen Mary University of London on the Mile End campus.

The bi-annual FOPARA workshop (http://fopara.cs.ru.nl) intends to serve as a forum for presenting original research results that are relevant to the analysis of resource (e.g., time, space, energy) consumption by computer programs. The workshop aims to bring together researchers who work on foundational issues with researchers who focus more on practical results. Therefore, both theoretical and practical contributions are encouraged. Also papers are encouraged that combine theory and practice. Typical topics are:

- Resource static analysis for embedded or/and critical systems
- Logical and machine-independent characterizations of complexity classes
- Logics closely related to complexity classes
- Type systems for controlling/inferring/checking complexity
- Semantic methods to analyze resources, including quasi-interpretations
- Practical applications of resource analysis
- Complexity analysis by term and graph rewriting

FOPARA 2015 was a two-phase workshop. All participants were invited to submit a draft paper describing the work to be presented at the workshop. These submissions were screened by the Program Committee chair to make sure they were within the scope of FOPARA and they appeared in the informal pre-proceedings that were available at the workshop. Submissions appearing in these draft proceedings were not peer-reviewed publications.

After the workshop, via an *open call* the opportunity was given to submit a paper to the formal proceedings, and authors of the draft proceedings including the invited speakers were in particular invited to submit a final paper incorporating the feedback from discussions at the workshop. These revised submissions were reviewed by the Program Committee using prevailing academic standards to select the best articles that will appear in the formal proceedings. The papers selected after the reviewing process are now published in this volume of the Springer LNCS series. The 2009, 2011, and 2013 FOPARA proceedings are published as LNCS Volumes 6324, 7177, and 8552, respectively.

Three FOPARA workshops preceded this event. The first edition of the FOPARA workshop was held in 2009 as part of Formal Methods Europe (FM2009) at the Technical University Eindhoven, The Netherlands. The second edition was co-located with Trends in Functional Programming (TFP2011) at Complutense University in Madrid, Spain. The third FOPARA workshop was co-located with WST 2013, the 13th

International Workshop on Termination, at the Bertinoro International Center for Informatics in Italy:

- FOPARA 2009: Eindhoven at Formal Methods Europe

  - http://resourceanalysis.cs.ru.nl/fopara/

- FOPARA 2011: Madrid at Trends in Functional Programming

  - http://dalila.sip.ucm.es/fopara11/

- FOPARA 2013: Bertinoro at Workshop on Termination

  - http://fopara2013.cs.unibo.it/

The 2015 FOPARA workshop was an inspiring event not only for the FOPARA community but also for the DICE community because the two workshop were co-located with a joint program. This may be a first step toward many future FOPARA-DICE joint program co-locations.

The 2015 edition of FOPARA was supported by the European Union via ICT COST Action IC1202: Timing Analysis on Code LEvel (TACLE).

August 2016   Marko van Eekelen
                                                        Ugo Dal Lago

Organization

Program Committee

Marko van Eekelen	Radboud University and Open University, The Netherlands
Ugo Dal Lago	University of Bologna, Italy
Elvira Albert	University Complutense Madrid, Spain
Josh Berdine	Microsoft Research, UK
Patrick Baillot	ENS-Lyon, France
Kerstin Eder	University of Bristol, UK
Clemens Grelck	University of Amsterdam, The Netherlands
Kevin Hammond	University of St. Andrews, UK
Manuel Hermenegildo	IMDEA, Madrid, Spain
Martin Hofmann	LMU, Munich, Germany
Thomas Jensen	IRISA, Rennes, France
Björn Lisper	Mälardalen University, Sweden
Hans-Wolfgang Loidl	Heriot-Watt University, UK
Kenneth MacKenzie	University of Glasgow, UK
Jean-Yves Marion	University of Lorraine, France
Greg Michaelson	Heriot-Watt University, UK
Georg Moser	University of Innsbruck, Austria
Romain Péchoux	University of Lorraine, France
Ricardo Peña	University Complutense of Madrid, Spain
Simona Ronchi della Rocca	University of Turin, Italy
Luca Roversi	University of Turin, Italy
Aleksy Schubert	Warsaw University, Poland
Simon Wegener	AbsInt Angewandte Informatik GmbH, Germany

Additional Reviewer

Bernard van Gastel	Open University and Radboud University

Contents

May-Happen-in-Parallel Analysis
with Condition Synchronization

Elvira Albert[1]([envelope]), Antonio Flores-Montoya[2], and Samir Genaim[1]

[1] Complutense University of Madrid (UCM), Madrid, Spain
elvira@sip.ucm.es
[2] Technische Universität Darmstadt (TUD), Darmstadt, Germany

Abstract. Concurrent programs can synchronize by means of conditions and/or message passing. In the former, processes communicate and synchronize by means of shared variables that several processes can read and write. In the latter, communication is by sending, receiving and waiting for messages. Condition synchronization is often more efficient but also more difficult to analyze and reason about. In this paper, we leverage an existing *may-happen-in-parallel* (MHP) analysis, which was designed for a particular form of message passing based on future variables, to handle condition synchronization effectively, thus enabling the analysis of programs that use both mechanisms. This is done by developing a *must-have-finished* analysis which is used to refine the MHP relations inferred by the original MHP analysis. The information inferred by an MHP has been proven to be essential to infer both safety properties (e.g., deadlock freedom) and liveness properties (termination and resource boundedness) of concurrent programs.

1 Introduction

With the trend of parallel and distributed systems and the emergence of multi-core architectures, the development of techniques and tools that help analyzing and verifying the behaviour of concurrent programs has become fundamental. Concurrent programs contain several processes (or tasks) that work together to perform a task. For that purpose, they communicate and synchronize with each other. Communication can be programmed using conditions or using *future variables* [8].

In order to develop our analysis, we consider a generic asynchronous language in which tasks can execute across different concurrent objects in parallel; and inside each object, tasks can interleave their computations. The language allows both synchronization using conditions and future variables. When future variables are used for synchronization, one process notifies through the future variable that its execution is completed and the notification is received by the process(es) waiting for its completion. In particular, the instruction Fut f=b!m(); posts an asynchronous task m on object b and allows the current task to synchronize with the completion of m by means of the future variable f. The instruction **await** f? is used to synchronize with the completion of m as follows. If the process

© Springer International Publishing Switzerland 2016
M. van Eekelen and U. Dal Lago (Eds.): FOPARA 2015, LNCS 9964, pp. 1–19, 2016.
DOI: 10.1007/978-3-319-46559-3_1

executing m has not finished when executing the **await**, the future f is not ready, and the current process is suspended. In such case, the processor can be released such that another pending process in the same object can take it and start to execute (thus interleaving its computation with the suspended task). When condition synchronization is used, one process writes into a variable that is read by another. Thus, instead of using future variables, the instruction **await** takes the form **await** b? where b is a Boolean condition involving shared variables. For instance, we can write **await** x!=**null** that synchronizes on the condition that the shared variable x is not **null**. The use of condition synchronization is known to pose challenges in static analysis. This is because it is difficult to automatically infer to which parts of the program the **await** instruction synchronizes. As a consequence, condition synchronization is more difficult to debug and analyze, while its main advantage is efficiency – it lacks the overhead of managing future variables. In contrast, future variable based synchronization is less efficient but it helps in producing concurrent applications in a less error-prone way as it is clearer how processes synchronize. The focus of the paper is to cover the more general **await** pred? construct, for an arbitrary predicate pred that can be a future variable or a boolean condition. The challenge is to retrieve precise information for this more general case, that lacks structural information inherent in the future-based synchronisation.

May-happen-in-parallel (MHP) is an analysis which identifies pairs of statements that can execute in parallel across several objects and in an interleaved way within an object (see [4,14]). MHP directly allows ensuring absence of data races in the access to the shared memory. Besides, it is a crucial analysis to later prove more complex properties like termination, resource consumption and deadlock freedom. In [10,16], MHP pairs are used to greatly improve the accuracy of deadlock analysis. The language we consider uses the instruction **f.get** to block the execution of the current task until the task associated with the future f has finished. Consider method "void m (Buffer b) {Fut f=b!foo(); f.get;}". Given two objects b_1 and b_2, if we execute the processes m(b_2) on b_1 and m(b_1) on b_2 in parallel, we can have a deadlock situation as they block waiting for each other to terminate. The MHP analysis allows discarding infeasible deadlocks when the instructions involved in a possible deadlock cycle cannot happen in parallel. For instance, if we are sure that the execution of m(b_2) in b_1 has finished before the execution of m(b_1) in b_2 starts (i.e., the tasks cannot happen in parallel), we can prove deadlock freedom. Also, MHP improves the accuracy of termination and cost analysis [5] since it allows discarding infeasible interleavings. For instance, consider a loop like "while (l!=null) {x=b!p(l.data); await x?; l=l.next;}", where the instruction **await** x? synchronizes with the completion of the asynchronous task to p. If the asynchronous task is not completed (x is not ready), the current task releases the processor and another task can take it. This loop terminates and has a bounded resource consumption provided no instruction that increases the length of the list l *interleaves* or *executes in parallel* with the body of this loop.

In this paper, we leverage an existing MHP analysis [4,14] developed for synchronization using future variables to handle condition synchronization effectively. Handling both future variables and shared memory synchronization is difficult because the analysis has to infer soundly the program points which the **await** instructions synchronize with and propagate them accordingly to both kinds of synchronization. Our analysis is based on the *must-have-finished* (MHF) property which allows us to determine that a given instruction will not be executed afterwards after a certain point. In particular, we aim at inferring MHF-sets which are the set of program points that MHF when the execution reaches a program point of interest. Developing our extension for condition synchronization amounts to developing an analysis that infers the required MHF-sets and using them to refine the original MHP analysis. We do this according to the following steps which constitute our original contributions:

1. We formally define the concrete notion of MHF, which assigns an MHF-set to each program point p and allows us to discard MHP pairs;
2. We define two auxiliary properties *must-have-happened* (expressed as MHH-sets for each program point) and *uniqueness* that combined, allow us to under-approximate the MHF-sets;
3. We define MHH-seeds, which are initial sets that under-approximate the MHH-sets for some selected program points. These seeds can be provided by the *user*, or automatically inferred. We propagate the MHH-seeds to other program points in the program in order to compute the MHH-sets;
4. We propose simple mechanisms to automatically infer MHF-seeds, and thus fully automatize the analysis; and
5. We discuss different ways in which MHF-sets allow eliminating MHP pairs.

We have implemented our analysis in the SACO system (Static Analyzer for Concurrent Objects) which can be used online through a web interface at costa.ls.fi.upm.es/saco. Our analysis has been evaluated on two industrial case studies developed by Fredhopper® which are available at the above site together with smaller programs. While previous analyses [10,11] required manual annotations to prove deadlock freedom of one case study, our MHP analysis for condition synchronization has allowed us to prove it in a fully automated way.

2 Language

We consider asynchronous programs with multiple task buffers (see [9]) which may concurrently interleave their computations. The concept of task buffer is materialized by means of an *object*. Tasks from different objects execute in parallel. Tasks can be synchronized with the completion of other tasks (from the same or a different object) using futures and conditions on shared variables. The number of objects does not have to be known a priori and objects can be dynamically created. Our model captures the essence of the concurrency and distribution models used in actor-languages (including concurrent objects [13], Erlang [1] and

Scala [12]) and in X10 [14], which rely on futures and message passing synchronization. It also has many similarities with the concurrency model in [9] which uses condition synchronization.

2.1 Syntax

A *program* consists of a set of classes, each of them can define a set of fields, and a set of methods. One of the methods, named main, corresponds to the initial method which is never posted or called and it is executing in an object with identifier 0. The grammar below describes the syntax of our programs. Here, T are types, m method names, e expressions, x can be field accesses or local variables, b is a Boolean condition involving fields, and y are future variables.

$$CL ::= class\ C\ \{\bar{T}\ \bar{f}; \bar{M}\} \qquad M ::= T\ m(\bar{T}\ \bar{x})\{s; \textbf{return}\ e; \}$$
$$s\ \ ::= s; s\ |\ x = e\ |\ \textbf{if}\ e\ \textbf{then}\ s\ \textbf{else}\ s\ |\ \textbf{while}\ e\ \textbf{do}\ s\ |$$
$$\textbf{await}\ y?\ |\textbf{await}\ \textbf{b}|\ x = \textbf{new}\ C(\bar{e})\ |\ y = x!m(\bar{e})\ |\ y.\textbf{get}$$

The notation $\bar{T}\ \bar{f}$ is used as a shorthand for the sequence $T_1\ f_1; \ldots; T_n\ f_n$, where T_i is a type and f_i is a field name. We use the special identifier this to denote the current object. For the sake of generality, the syntax of expressions and conditions is left free and also the set of types is not specified. Note that every method ends with a "**return** e" (if the return type is void we use "**return** 0").
    Each object has a heap with the values assigned to its fields. The concurrency model is as follows. Each object has a lock that is shared by all tasks that belong to the object. Data synchronization is by means of future variables and conditions as follows. An **await** y? instruction is used to synchronize with the result of executing task $y = b!m(\bar{z})$ such that **await** y? is executed only when the future variable y is available (and hence the task executing m on object b is finished). In the meantime, the object's lock can be released and some *pending* task on that object can take it. The synchronization instruction **await** b blocks the execution until the Boolean condition b evaluates to *true*, and allows other tasks to execute in the meantime. The instruction $y.\textbf{get}$ blocks the object (no other task of the same object can run) until y is available, i.e., the execution of $m(\bar{z})$ on b is *finished*. The difference from **await** y? is in that is *blocks* the object.
    Note that our concurrency model is *cooperative* as processor release points are explicit in the code, in contrast to a *preemptive* model in which a higher priority task can interrupt the execution of a lower priority task at any point. Without loss of generality, we assume that all methods in a program have different names.

2.2 Semantics

A *program state* $St = \textbf{Obj}$ is of the form $\textbf{Obj} \equiv obj_1\ \|\ \ldots\ \|\ obj_n$ denoting the parallel execution of the created objects. Each *object* is a term $obj(oid, f, lktid, \mathcal{Q})$ where *oid* is the object identifier, f is a mapping from the object fields to their values, *lktid* is the identifier of the *active task* that holds the object's lock or $\bot$ if the object's lock is free, and $\mathcal{Q}$ is the set of tasks in the object. Only one task can be *active* (running) in each object and has its *lock*. All other tasks are *pending* to be executed, or *finished* if they terminated and released the lock. A *task* is a term $tsk(tid, m, l, s)$ where tid is a unique task identifier, m is the method name executing in the task, l is a mapping from local (possibly future) variables to their values, and s is the sequence of instructions to be executed or $s = \epsilon(v)$ if the task has terminated and the return value v is available. Each instruction is associated to a program point in which the instruction appears in the program. By abuse of notation, we sometimes use program points rather than the instructions in the sequence s and in the examples. Created objects and tasks never disappear from the state.

$$\frac{fresh(oid'),\ l' = l[x \to oid'],\ t = tsk(tid, m, l, \langle x = \textbf{new}\ C(\bar{e}); s \rangle), f' = init\_atts(C, \bar{e})}{obj(oid, f, tid, \{t\} \cup \mathcal{Q}) \parallel B \rightsquigarrow}$$
$$\text{(NEWOBJECT)} \qquad obj(oid, f, tid, \{tsk(tid, m, l', s)\} \cup \mathcal{Q}) \parallel obj(oid', f', \bot, \{\}) \parallel B$$

$$\frac{t = tsk(tid, \_, \_, s) \in \mathcal{Q},\ s \neq \epsilon(v)}{\text{(SELECT)} \quad obj(oid, f, \bot, \mathcal{Q}) \parallel B \rightsquigarrow obj(oid, f, tid, \mathcal{Q}) \parallel B}$$

$$\frac{l(x) = oid_1,\ fresh(tid_1),\ l' = l[y \to tid_1],\ l_1 = buildLocals(\bar{z}, m_1)}{obj(oid, f, tid, \{tsk(tid, m, l, \langle y = x!m_1(\bar{z}); s \rangle)\} \cup \mathcal{Q}) \parallel obj(oid_1, f_1, \_, \mathcal{Q}') \parallel B \rightsquigarrow}$$
$$\text{(ASYNC)} \qquad obj(oid, f, tid, \{tsk(tid, m, l', s)\} \cup \mathcal{Q}) \parallel$$
$$obj(oid_1, f_1, \_, \{tsk(tid_1, m_1, l_1, body(m_1))\} \cup \mathcal{Q}') \parallel B$$

$$\frac{c = y?,\ l(y) = tid_1,\ tsk(tid_1, \_, \_, s_1) \in \mathcal{Q}, s_1 = \epsilon(v) \vee c = b,\ eval(b, f, l) = true}{obj(oid, f, tid, \{tsk(tid, m, l, \langle \textbf{await}\ c; s \rangle)\} \cup \mathcal{Q}) \parallel B \rightsquigarrow}$$
$$\text{(AWAIT1)} \qquad obj(oid, f, tid, \{tsk(tid, m, l, s)\} \cup \mathcal{Q}) \parallel B$$

$$\frac{c = y?,\ l(y) = tid_1,\ tsk(tid_1, \_, \_, s_1) \in \mathcal{Q}, s_1 \neq \epsilon(v) \vee c = b,\ eval(b, f, l) = false}{obj(oid, f, tid, \{tsk(tid, m, l, \langle \textbf{await}\ c; s \rangle)\} \cup \mathcal{Q}) \parallel B \rightsquigarrow}$$
$$\text{(AWAIT2)} \qquad obj(oid, f, \bot, \{tsk(tid, m, l, \langle \textbf{await}\ c; s \rangle)\} \cup \mathcal{Q}) \parallel B$$

$$\frac{l(y) = tid_1,\ tsk(tid_1, \_, \_, s_1) \in \textsf{Obj}, s_1 = \epsilon(v), l' = l[x \to v]}{\text{(GET)} \quad obj(oid, f, tid, \{tsk(tid, m, oid, l, \langle x = y.\textbf{get}; s \rangle)\} \cup \mathcal{Q}) \parallel B \rightsquigarrow}$$
$$obj(oid, f, tid, \{tsk(tid, m, oid, l', s)\} \cup \mathcal{Q}) \parallel B$$

$$\frac{v = l(x)}{\text{(RETURN)} \quad obj(oid, f, tid, \{tsk(tid, m, l, \langle \textbf{return}\ x; \rangle)\} \cup \mathcal{Q}) \parallel B \rightsquigarrow}$$
$$obj(oid, f, \bot, \{tsk(tid, m, l, \epsilon(v))\} \cup \mathcal{Q}) \parallel B$$

Fig. 1. Summarized semantics

The execution of a program starts from an initial state where we have an initial object with identifier 0 which has no fields and is executing task 0 of the form $S_0 = obj(0, [], 0, \{tsk(0, \text{main}, l, body(\text{main}))\})$. Here, l maps parameters to their initial values and local reference and future variables to **null** (standard initialization), and $body(m)$ refers to the sequence of instructions in the method m. The execution proceeds from S_0 by selecting *non-deterministically* one of the objects and applying the semantic rules depicted in Fig. 1. We omit the treatment of the sequential instructions as it is standard.

NEWOBJECT: an active task *tid* in object *oid* creates an object oid' of type C, its fields are initialized (init_atts) and oid' is introduced to the state with a free lock. SELECT: this rule selects non-deterministically one of the tasks that is in queue and is not finished, and it obtains its object's lock. ASYNC: A method call creates a new task (the initial state is created by *buildLocals*) with a fresh task identifier tid_1 which is associated to the corresponding future variable y in l'. We have assumed that $oid \neq oid_1$, but the case $oid = oid_1$ is analogous, the new task tid_1 is simply added to Q of oid. AWAIT1: It deals with synchronization both on future variables and shared memory. If the future variable we are awaiting for points to a finished task or the condition evaluates to **true** (we use function *eval* to evaluate the condition), the **await** can be completed. When using future variables, the finished task t_1 is looked up in all objects in the current state (denoted Obj). Similarly, the evaluation of the condition will require accessing object fields and possibly local variables (thus f and l are looked up). AWAIT2: Otherwise, the task yields the lock so that any other task of the same object can take it. GET: It waits for the future variable but without yielding the lock. Then, it retrieves the value associated with the future variable. RETURN: When **return** is executed, the return value is stored in v so that it can be obtained by the future variable that points to that task. Besides, the lock is released and will never be taken again by that task. Consequently, that task is *finished* (marked by adding the instruction $\epsilon(v)$) but it does not disappear from the state.

```
1  void main(int m, int n) {        21  void incrf() {
2    A a = new A(m,n);              22     f=2*f;
3    a!init();                       23     return 0;
4    a!go();                         24  }
5    return 0;                       25
6  }                                 26  void incrg() {
7                                    27     g=2*g;
8  class A <int g, int f>{           28     return 0;
9    C x = null;                     29  }
10                                   30
11   void init() {                   31  void go() {
12     while (g>0) {                 32     await x!=null;
13       Fut y = this!incrf();       33     while (f>0) {
14       await y?                    34       Fut z = this!incrg();
15       g--;                        35       await z?
16     }                             36       f--;
17     x = new C();                  37     }
18     return 0;                     38     return 0;
19   }                              39  }
20                                  40  }
```

Fig. 2. Synchronization using Boolean conditions.

Example 1. Our motivating example is shown in Fig. 2. It consists of two classes A and C and a main method from which the execution starts. The main method receives two input parameters. For simplicity class C and return instructions have been omitted. We will see later that a naïve analysis of the synchronization instructions will report termination and resource boundness problems that are spurious (i.e., false alarm). The trace $S_0 \rightsquigarrow S_1 \rightsquigarrow^* S_2 \rightsquigarrow^* S_3 \rightsquigarrow S_4$ where

$S_0 \equiv obj(0, [], 0, \{tsk(0, \mathsf{main}, l_0 = [m \mapsto m_0, n \mapsto n_0], \tilde{2}))\})$
$S_1 \equiv obj(0, [], 0, \{tsk(0, \mathsf{main}, l_0, \tilde{3})\}) \parallel obj(1, f_1, \bot, \{\})$
$S_2 \equiv obj(0, [], 0, \{tsk(0, \mathsf{main}, l_0, \tilde{5})\}) \parallel obj(1, f_1, \bot, \{tsk(1, \mathsf{init}, l_1, \tilde{12}), \{tsk(2, \mathsf{go}, l_2, \tilde{32})\})$
$S_3 \equiv obj(0, [], 0, \{tsk(0, \mathsf{main}, l_0, \tilde{5})\}) \parallel obj(1, f_1, 2, \{tsk(1, \mathsf{init}, l_1, \tilde{12}), \{tsk(2, \mathsf{go}, l_2, \tilde{32})\})$
$S_4 \equiv obj(0, [], 0, \{tsk(0, \mathsf{main}, l_0, \tilde{5})\}) \parallel obj(1, f_1, \bot, \{tsk(1, \mathsf{init}, l_1, \tilde{12}), \{tsk(2, \mathsf{go}, l_2, \tilde{32})\})$
$S_5 \equiv obj(0, [], 0, \{tsk(0, \mathsf{main}, l_0, \tilde{5})\}) \parallel obj(1, f_1, 1, \{tsk(1, \mathsf{init}, l_1, \tilde{12}), \{tsk(2, \mathsf{go}, l_2, \tilde{32})\})$

corresponds to few execution steps starting from method main and two constant values m_0 and n_0 as input parameters. (1) S_0 is the initial state, it includes one object with the main method whose next instruction to be executed is the one at program point 2 (denoted as $\tilde{2}$) and the local variable mapping l_0 keeps the bindings for the input arguments; (2) S_1 is obtained from S_0 by executing the first instruction of method main, using rule NEWOBJECT, that creates object 1 and initializes the two class fields f and g with the values that are passed as arguments in the **new** instruction (the bindings for the fields are kept in f_1); (3) S_2 is obtained from S_1 by applying rule ASYNC twice, to execute program points 3 and 4 of the main method to create the tasks 1 and 2 and add them to the queue of object 1; (4) S_3 is obtained from S_2 by applying rule SELECT that selects task 2 from object 1 to start to execute (5) S_4 is obtained from S_3 by applying rule AWAIT2 that evaluates the boolean condition of the **await** at 32 to **false** and thus releases the processor. Note that this condition synchronizes with the instruction at line 17 which creates the object x and thus allows task 2 to move forward; (6) at S_5 task 1 is selected for execution at object 1 and its execution can proceed till the loop finishes iterating and afterwards object x is created. Up to that point, the task 2 is blocked in the **await** instruction.

3 MHP: Concrete Definition and Static Analysis

In this section, we define the property may-happen-in-parallel, and summarize the main points of the analysis of [4] which over-approximates this property. Finally, we describe a straightforward extension of this analysis to handle condition synchronization which is simple but imprecise. The extension will cover the more general: *await pred?* construct, for an arbitrary predicate *pred* that includes both future variables and boolean conditions.

3.1 Concrete Definition

We first define the concrete property "MHP" that we want to approximate using static analysis. In what follows, we assume that instructions are labelled such that it is possible to obtain the corresponding program point identifiers. Given a sequence of instructions s, we use $pp(s)$ to refer to the program point identifier associated with its first instruction, and $pp(\epsilon(v))$ refers to the exit program point.

Definition 1. *A program point p is active in a state $S = \texttt{Obj}$ within task tid, iff there is $obj(oid, \_, \_, Q) \in \texttt{Obj}$ and $tsk(tid, \_, \_, s) \in Q$ such that $pp(s) = p$.*

We sometimes say that p is active in task S without referring to the corresponding task identifier. Intuitively, this means that there is a task in S whose next instruction to be executed is the one at program point p.

Definition 2. *Given a program P, its MHP is defined as $\mathcal{E}_P = \cup \{\mathcal{E}_S | S_0 \rightsquigarrow^* S\}$ where $\mathcal{E}_S = \{(p_1, p_2) \mid p_1$ is active in S within tid_1, p_2 is active in S within tid_2 and $tid_1 \neq tid_2\}$.*

The above definition considers the union of the pairs obtained from all derivations from S_0. This is because execution is non-deterministic in two dimensions: (1) in the selection of the object that is chosen for execution, different behaviours (and thus MHP pairs) can be obtained depending on the execution order, and (2) when there is more than one task, the selection is non-deterministic.

The MHP pairs can originate from *direct* or *indirect* task creation relationships. For instance, in the program of Fig. 2, the parallelism between the points of the tasks executing init and go is *indirect* because they do not invoke one to the other directly, but a third task main invokes both of them. However, the parallelism between the points of the task go and those of incrg is *direct* because the first one invokes directly the latter one. Definition 2 captures both forms of direct and indirect parallelism and they are indistinguishable in the MHP pairs.

3.2 Static Analysis

In [4], an MHP static analysis is presented. Intuitively, the analysis is formalized as a two-phase process: (1) First, methods are locally analyzed and the analysis learns about the tasks invoked within each method, the instructions **await**, **get** and **return** in the following way. A method call creates a task that may run in parallel with the current task. For instance, the task created at 3 runs in parallel with the current one. If the object in which the task is posted is **this** (e.g., when we invoke incrf()) then the task is pending in the object **this** since we know that it cannot start to execute until a release point is reached in the current task. From **await** and **get** used with futures, the analysis learns that the corresponding tasks have finished after such instructions execute. For instance, at 35, we know that the task created at 34 is finished. Also, at 14 we know that the task created at 13 is finished. This is an essential piece of information for precision. Also, tasks that were pending in this object may become active at the **await** point. For instance, the call considered pending before incrf() would be considered as potentially active at 14, since the processor might be released at this point (and indeed it will necessarily as y cannot be ready until incrf terminates). Likewise, from **return**, we learn that tasks that were pending might become active, as the processor is released by the current task. (2) Second, the information gathered at the local phase is composed by means of an MHP-graph, denoted $\mathcal{G}_P$, which contains all MHP information for a program P. The graph allows us to obtain the (abstract) set of MHP pairs $\tilde{\mathcal{E}}_P$, which are an over-approximation of $\mathcal{E}_P$.

Informally, the set $\tilde{\mathcal{E}}_P$ is obtained by checking certain reachability conditions on $\mathcal{G}_P$. Such conditions are not relevant to the contents of this paper (see [4]).

Theorem 1 (soundness [4]). $\mathcal{E}_P \subseteq \tilde{\mathcal{E}}_P$.

3.3 Naïve Extension

The analysis above only handles synchronization instructions, "**await** y?" and "y.**get**", that allow synchronizing one task with the completion of other tasks. Handling "**await** b", where b is a Boolean condition, poses new challenges on the MHP analysis since, unlike "**await** y?", it is not immediate to know which part of the program it synchronizes with, i.e., which instructions make this condition satisfiable. The analysis in Sect. 3.2 can be naïvely extended to soundly, but not precisely, handle this new instruction. This is simply done by treating it, in the first phase of the analysis, as "**await** z?" where z is a fresh variable that does not appear anywhere in the program. This allows pending tasks to become active after the **await**, and thus guarantees soundness. However, it will not mark any task as finished as z is not associated to any instruction or task, and thus it leads to imprecise results. It should be noted that making such replacement would not lead to an equivalent program, as the **await** z? would block the task forever. However, the analysis of the program after such replacement is sound because it considers all possibilities and, in particular, the one in which the **await** z? succeeds.

Example 2. Consider the program depicted in Fig. 2, and note that program point 32 synchronizes with program point 17 using a Boolean condition. Applying the basic MHP analysis with the above naïve extension, we will conclude that $(15, 33)$ and $(36, 12)$ are MHP pairs. These are spurious pairs since go will not proceed to program point 33 until program point 17 has been executed, i.e., go waits for init to finish at program point 17. In practice, these spurious MHP pairs lead to imprecision in analyses for more complex properties that rely on this MHP information. This is the case of the termination and resource analyses of [5] which reports a false alarm. Non-termination (and resource unboundedness) is actually feasible if we omit the synchronization condition at line 17, and is obtained as follows: (1) in main we invoke init and go; (2) go starts to execute and calls incrg that executes and increases the counter g by one; (3) at the release point in line 35, task init starts to execute and invokes init (which increases the counter of the loop at go by multiplying it by two); (4) thus, the execution of both loops interleaves and the tasks invoked within the loops modify the counter of the other loop such that the interleaved execution will not terminate; In the presence of the condition at line 32 this termination and resource boundness problem becomes infeasible, since the loop of go cannot start until init has finished and all instances of the tasks init that modify the loop counter and thus finished as well. It will be eliminated later when the MHP analysis is enhanced to infer that (15 and 33), and also (36 and 12) cannot execute in parallel.

4   MHP Analysis with Condition Synchronization

The goal of this section is to leverage the MHP analysis described in Sect. 3 to handle condition synchronization. For this, we assume that the basic MHP analysis (with the näive extension of Sect. 3.3) has been applied, and the result is given as a set of MHP pairs $\tilde{\mathcal{E}}_P$. Then, we present an extension that eliminates infeasible MHP pairs, as the one of Example 2, from this set. This extension is based on a property that we call *must-have-finished* that can be directly used to eliminate infeasible MHP pairs.

The rest of this section is organized as follows: Sect. 4.1 defines the *must-have-finished* property and discusses a specific way to under-approximate it; Sects. 4.2, 4.3, 4.4 describe a practical way to carry out this under-approximation; and finally Sect. 4.5 sketches some important improvements to the overall analysis.

4.1   Must-Have-Finished Property

The *must-have-finished* (MHF) property is defined by means of MHF-sets that are assigned to each program point. Intuitively, the MHF-set for a given program point p is a set of program points that will never execute after p has been reached. Next we formally define the MHF property.

Definition 3. *Given two program points p_1 and p_2, we say that p_2 is in the MHF-set of p_1, denoted $p_2 \in \mathtt{mhfset}(p_1)$, iff for any trace $t \equiv S_0 \leadsto^* S \leadsto^* S'$, if p_1 is active in S, then p_2 is not active in S'.*

Intuitively, this definition states that whenever p_1 is reached (in state S), then p_2 is not active from that state on (note that S' could be equal to S). Clearly, if $p_2 \in \mathtt{mhfset}(p_1)$ then p_1 and p_2 cannot execute in parallel.

Corollary 1. *If $p_2 \in \mathtt{mhfset}(p_1)$, then $(p_1, p_2) \notin \mathcal{E}_P$.*

Example 3. For the program of Fig. 2, we have $17 \in \mathtt{mhfset}(33)$. This is because for method go to proceed to program point 33, the condition at program point 32 must be satisfied. The instruction at program point 17 is the only one that makes this condition satisfiable and it is executed only once. Therefore, it will not execute again after 33 has been executed. This will be used in Sect. 4.5 to guarantee that the method init, together with all instances of method incrg are finished at program point 33.

MHF is a non-trivial undecidable property, and thus we aim at computing under-approximations of it, that can be soundly used in Corollary 1, as follows: in order to include p_2 in the MHF-set of p_1 we require that (1) p_2 *must-have-happened* (MHH) before p_1, i.e., whenever p_1 is reached, then p_2 must have been executed in a previous transition at least once; and (2) p_2 is *unique*, i.e., p_2 executes at most once in any trace (This property will be later relaxed in Sect. 4.5). These two properties together imply $p_2 \in \mathtt{mhfset}(p_1)$. Next we formally define the notion of MHH-sets and the set of unique program points, and then explain how they are combined to under-approximate the MHF-sets.

Given a trace $t \equiv S_0 \leadsto^* S_n$, we say that program point p is *executed* in step $i < n$ of t, iff p is active in S_i within task tid and it is not active in S_{i+1} within task tid (note that p could be active in S_{i+1} within $tid' \neq tid$).

Definition 4. *We say that p_2 is in the MHH-set of p_1, denoted $p_2 \in \text{mhhset}(p_1)$, iff for any trace $t \equiv S_0 \leadsto^* S_n$, if p_1 is active in S_n then p_2 is executed at least once in step $i < n$ of t.*

Given a trace $t \equiv S_0 \leadsto^* S_n$, we say that a program point p is *reached* in state S_i, with $0 \leq i \leq n$, iff p is active in S_i within task tid, and, either $i = 0$ or p is not active in S_{i-1} within task tid. This means that task tid has reached program point p due to the execution step $S_{i-1} \leadsto S_i$, or, when $i = 0$, p is active in the initial state S_0 within task tid.

Definition 5. *A program point p is unique iff in any trace it is reached at most once. The set of unique program points is denoted by $\mathcal{U}$.*

Lemma 1. $\text{mhhset}(p) \cap \mathcal{U} \subseteq \text{mhfset}(p)$

Proof. Let p' be a program point such that (1) $p' \in \text{mhhset}(p)$ and (2) $p' \in uniqueSet$. For any trace $t \equiv S_0 \leadsto^* S_n \leadsto \ldots$ where p is active in S_n. Because of (1) there exists a $S_i \leadsto S_{i+1}$ $i < n$ where p' is executed (p' is active in S_i and it is not active in S_{i+1}). Because of (2) p' becomes active at most once in t (in S_i). Therefore, p' is not active for any S_j $j > n$ and $p' \in \text{mhfset}(p)$ □

Unfortunately the MHH and uniqueness properties are still non-trivial. In Sects. 4.2 and 4.3 we describe an analysis for under-approximating the MHH-sets, and in Sect. 4.4 we explain how we under-approximate the set $\mathcal{U}$.

It is worth noting that Definition 3 and the corresponding under-approximation, capture behaviours in which one part of the program synchronizes with some initialization tasks (since we forbid p_2 to execute again after p_1). This might seem restrictive since it is common to have several instances of the same initialization method, for different objects, executing independently at different times. However, in practice, our analysis uses object-sensitive information [3,15] and thus considers instances of the same method, when they run on different objects, as if they were different methods. The above formalism can be trivially adapted to an object-sensitive setting, but for the sake of simplifying the presentation, we keep it object-insensitive while our implementation is object-sensitive.

Example 4. Consider the program of Fig. 2, and suppose that the code of main is duplicated, e.g., add "A b = **new** A(); b!init(); b!go();" immediately after line 4. In such case, program point 17 must execute at least once before program point 33. However, since there are two instances of init, one will have to execute before program point 33, but the other might start executing after one of the instances of method go has reached program point 33. Ignoring the object-sensitive information we would conclude that $17 \notin \text{mhfset}(33)$. This is indeed correct if a and b were pointing to the same objects, which is not the case.

By taking object-sensitive information into account, our implementation considers the two calls to init (resp. go) as calls to different methods, and thus it is able to prove that $17 \in \text{mhfset}(33)$ for objects a and b.

4.2 Under-Approximating MHH-sets

Our analysis for inferring MHH-sets consists of the following two steps: in the first one we extract what we call MHH-seeds, which are under-approximations of the MHH-sets of some selected program points; and then, in a second step, these seeds are propagated to other program points. In this section we detail the second step, while for the first step we just define the notion of MHH-seeds and leave the automatic inference details to Sect. 4.3.

Definition 6 (MHH-seeds). *An MHH*-seed *for a program point p, denoted* mhhseed(p), *is a subset of* mhhset(p).

As the above definition indicates, the MHH-seed of a program point p is simply a subset of mhhset(p). However, these seeds correspond to elements of mhhset(p) that are relatively easy to compute (see Sect. 4.3). For now, we assume that every program point is assigned an MHH-seed. Note that, in general, the MHH-seeds of most program points is the empty set. MHH-seeds can be provided by the user, or automatically inferred as discussed in Sect. 4.3.

Example 5. For the program in Fig. 2, we let: mhhseed$(33) = \{17, 18\}$, due to the synchronization of program points 17 and 32; mhhseed$(15) = \{23\}$, because program point 14 waits for init to finish; and mhhseed$(36) = \{28\}$ since program point 35 waits for incrg to finish. For any other p we let mhhseed$(p) = \emptyset$.

Next we describe how to propagate the MHH-seeds to other program points, which is done using a *must* data-flow analysis [18]. We let P_p be the set of all program points of a program P. Our analysis domain is the complete lattice induced by the partial order $D = \langle \wp(P_p), \subseteq \rangle$ where $\wp(P_p)$ is the set of all sets of program points. Note that the least upper-bound and the greatest lower-bound are $\cup$ and $\cap$ respectively, and that the bottom and top elements are $\emptyset$ and P_p respectively. We let pre(p) be the set of program points that immediately precede program point p in the control flow. When p is an entry program point of method m, then pre(p) consists of the program points at which m is invoked.

Let us first explain the intuition behind the data-flow equations that we generate below. Let E_p stands for an MHH-set at program point p. Intuitively, we can construct E_p as follows: (1) let $E_{p'}$ be the MHH-set of $p' \in$ pre(p), then we can include $\cap\{E_{p'} \cup \{p'\} \mid p' \in$ pre$(p)\}$ in E_p, i.e., we take the intersection of all MHH-sets of program points that precede p; and (2) for each $p' \in$ mhhseed(p), we can include $E_{p'} \cup \{p'\}$ in E_p, i.e., p' is included in E_p since by definition it is in the MHH-set of p, but we can also add the points in the MHH-set of p' to that of p. Next we define the set of data-flow equations that formalizes the above intuition.

Definition 7 (MHH equations). *Given a program P, its system of MHH equations E_P includes an equation $E_p = E_p^1 \cup E_p^2$, for each program point p, where $E_p^1 = \cap\{E_{p'} \cup \{p'\} \mid p' \in$ pre$(p)\}$ and $E_p^2 = \cup\{E_{p'} \cup \{p'\} \mid p' \in$ mhhseed$(p)\}$.*

Note that E_p^1 takes the intersection of the MHH-sets of all program points that reach p in one step. This is because it is a *must* analysis and thus we should take the information that holds on all paths that lead to p. On the other hand, E_p^2 takes the union of the MHH-sets of those program points that are guaranteed to have been executed before executing p (i.e., its MHH-seeds).

Computing the MHH-sets amounts to finding the *greatest solution* for E_P, i.e., find a value for each E_p such that all equations in E_P are satisfied. In what follows, such a solution is denoted by $\tilde{E}_P : P_p \mapsto D$. Computing $\tilde{E}_P$ is a quite standard process [18], and a simple algorithm can be as follow. We start with an initial solution $\tilde{E}_P$ in which each $\tilde{E}_P(p)$ is assigned the top element P_p, and then we refine it by applying the following steps repeatedly until a fixed-point is reached: (i) substitute the current solution in the right-hand side of the equations E_P, and obtain a new solution $\tilde{E}'_P$; and (ii) set each $\tilde{E}_P(p)$ to $\tilde{E}_P(p) \cap \tilde{E}'_P(p)$.

Example 6. The following is the set of MHH equations for the program of Fig. 2:

$$
\begin{aligned}
E_2 &= \emptyset \\
E_3 &= E_2 \cup \{2\} \\
E_4 &= E_3 \cup \{3\} \\
E_5 &= E_4 \cup \{4\} \\
E_{12} &= (E_3 \cup \{3\}) \cap \\
       &\quad (E_{15} \cup \{15\}) \\
E_{13} &= E_{12} \cup \{12\} \\
E_{14} &= E_{13} \cup \{13\}
\end{aligned}
\qquad
\begin{aligned}
E_{15} &= E_{14} \cup \{14\} \cup \\
E_{23} &\quad \cup \{23\} \\
E_{17} &= E_{12} \cup \{12\} \\
E_{18} &= E_{17} \cup \{17\} \\
E_{22} &= E_{13} \cup \{13\} \\
E_{23} &= E_{22} \cup \{22\} \\
E_{27} &= E_{34} \cup \{34\} \\
E_{28} &= E_{27} \cup \{27\}
\end{aligned}
\qquad
\begin{aligned}
E_{32} &= E_4 \cup \{4\} \\
E_{33} &= ((E_{32} \cup \{32\}) \cap (E_{36} \cup \{36\})) \\
       &\quad \cup E_{17} \cup \{17\} \cup E_{18} \cup \{18\} \\
E_{34} &= E_{33} \cup \{33\} \\
E_{35} &= E_{34} \cup \{34\} \\
E_{36} &= E_{35} \cup \{35\} \cup E_{28} \cup \{28\} \\
E_{38} &= E_{33} \cup \{33\}
\end{aligned}
$$

Solving these equations using the algorithm described results in the following values of $\tilde{E}_P$ in the different iterations:

	#1	#2	#3	#4	#5	#6	$\cdots$	#9
$\tilde{E}_P(2)$	{}	{}	{}	{}	{}	{}		{}
$\tilde{E}_P(3)$	P_p	{2}	{2}	{2}	{2}	{2}		{2}
$\tilde{E}_P(4)$	P_p	P_p	{2,3}	{2,3}	{2,3}	{2,3}		{2,3}
$\tilde{E}_P(5)$	P_p	P_p	P_p	{2,3,4}	{2,3,4}	{2,3,4}		{2,3,4}
$\tilde{E}_P(12)$	P_p	P_p	{2,3}	{2,3}	{2,3}	{2,3}		{2,3}
$\tilde{E}_P(13)$	P_p	P_p	P_p	{2,3,12}	{2,3,12}	{2,3,12}		{2,3,12}
$\tilde{E}_P(14)$	P_p	P_p	P_p	P_p	{2,3,12,13}	{2,3,12,13}		{2,3,12,13}
$\tilde{E}_P(15)$	P_p	P_p	P_p	P_p	P_p	P_p		{2,3,12,13,22}
$\tilde{E}_P(17)$	P_p	P_p	P_p	{2,3,12}	{2,3,12}	{2,3,12}		{2,3,12}
$\tilde{E}_P(18)$	P_p	P_p	P_p	P_p	{2,3,12,17}	{2,3,12,17}		{2,3,12,17}
$\tilde{E}_P(22)$	P_p	P_p	P_p	P_p	{2,3,12,13}	{2,3,12,13}	$\cdots$	{2,3,12,13}
$\tilde{E}_P(23)$	P_p	P_p	P_p	P_p	P_p	{2,3,12,13,22}		{2,3,12,13,22}
$\tilde{E}_P(27)$	P_p	P_p	P_p	P_p	P_p	P_p		{2,3,4,12,17,18,32,33,34}
$\tilde{E}_P(28)$	P_p	P_p	P_p	P_p	P_p	P_p		{2,3,4,12,17,18,27,32,33,34}
$\tilde{E}_P(32)$	P_p	P_p	P_p	{2,3,4}	{2,3,4}	{2,3,4}		{2,3,4}
$\tilde{E}_P(33)$	P_p	P_p	P_p	P_p	P_p	{2,3,4,12,17,18,32}		{2,3,4,12,17,18,32}
$\tilde{E}_P(34)$	P_p	P_p	P_p	P_p	P_p	P_p		{2,3,4,12,17,18,32,33}
$\tilde{E}_P(35)$	P_p	P_p	P_p	P_p	P_p	P_p		{2,3,4,12,17,18,32,33,34}
$\tilde{E}_P(36)$	P_p	P_p	P_p	P_p	P_p	P_p		{2,3,4,12,17,18,27,28,32−35}
$\tilde{E}_P(38)$	P_p	P_p	P_p	P_p	P_p	P_p		{2,3,4,12,17,18,32,33}

Lemma 2. *Let $\tilde{E}_P$ be the greatest solution for E_P, then for all $p \in P_p$ we have $\tilde{E}_P(p) \subseteq \mathtt{mhhset}(p)$.*

Proof. (sketch) We assume that the fixpoint algorithm described above is used compute the greatest solution $\tilde{E}_P$. Without loss of generality, we assume that $\mathtt{pre}(p) = \emptyset$ for the entry program point p of method main. In what follows the length of a trace is computed as the number of its states. Let $\tilde{E}_P^i$ be the solution at iteration $i \geq 1$ of the fixpoint algorithm, then we claim that: if $p' \notin \mathtt{mhhset}(p)$ when considering all traces of length at most $i \geq 1$, then $p' \notin \tilde{E}_P^i(p)$. Clearly this claim implies the above lemma since, in addition, the algorithm terminates because the underlying domain does not include infinite chains.

The above claim can be proved by induction on i. The base-case it straightforward, since from traces of length at most 1 we can only conclude that the MHH-set of the entry program point p of method main is $\emptyset$, and, we have an equation $E_p = \emptyset$. For the induction step, we assume that the claim holds for some $i \geq 0$ and show that it holds for $i + 1$ as well. Observe that if we can conclude that $p' \notin \mathtt{mhhset}(p)$ when considering traces of length at most $i + 1$ but not when considering traces of length at most i, then there must be a program point $p'' \in \mathtt{pre}(p)$ for which $p' \notin \mathtt{mhhset}(p'')$ can be concluded using traces of length at most i. Thus, by the induction hypotheses, we have $p' \notin \tilde{E}_P^i(p'')$. Now since the definition of E_p takes the intersection of all its predecessors, we get $p' \notin \tilde{E}_P^{i+1}(p')$. Note that the propagation of seeds through E_p^2 does not interfere with the above reasoning since the MHH-seeds are assumed to be correct.   □

4.3   Automatic Inference of MHH-Seeds

In this section we explain how we automatically infer the MHH-seeds that are required in Sect. 4.2. Note that MHH-seeds are required only for program points with the instructions "**await** y?" (or equivalently "$y.\mathbf{get}$") or "**await** b", since these instructions are the ones used for synchronization. For any other program point p we set $\mathtt{mhhseed}(p)$ to $\emptyset$.

In what follows, for the sake of simplifying the presentation, we assume that **await** instructions do not appear at the end of a conditional branch or at the end of a loop body; otherwise, we can add a special **skip** instruction to avoid such cases. Recall that in order to include p_2 in $\mathtt{mhhseed}(p_1)$ we should guarantee that whenever p_1 is executed, p_2 must have executed before at least once.

Extracting MHH-seeds from "await y?" instructions. Future variables are pointers to tasks, i.e., when a method is invoked a corresponding task is created and a future variable is assigned the identifier (pointer) of this task. Thus, we could use points-to analysis to understand to which *methods* future y refers in a given instruction "**await** y?". In practice, we use the points-to analysis as described in [10] which infers, for each "**await** y?", a *set of methods* that correspond to the tasks that y might point-to. Clearly, if it is guaranteed that y refers to only one method m, we can add the program point of the **return** instruction of m to the MHH-set of the program point p' that immediately follows the program point p of the instruction "**await** y?".

Example 7. Consider the program of Fig. 2. Applying this technique to the **await** instruction at program point 35, we add program point 28 to mhhseed(36) since the future variable **z** refers to method incrg.

Extracting MHH-seeds from "await b" instructions. Recall that in Sect. 2 we left the syntax of the Boolean conditions free, and, in practice, it can be any Boolean expression that does not include method calls. However, we have to restrict ourselves to some specific forms since our method for extracting the MHH-seeds of "**await** b" is based on seeking instructions that make b satisfiable. The MHH-seed of any "**await** b" that does not fall into these forms will be $\emptyset$.

The general pattern of our method for extracting the MHH-seeds of "**await** b" is as follows: (1) we first look for an instruction that makes the condition b satisfiable, say we found one at program point p; (2) we prove that there are no other instructions that make b satisfiable or unsatisfiable — the only allowed way to be unsatisfiable is by the default values assigned to the variables when creating the corresponding objects; and (3) we define the MHH-seeds of the instruction that follows "**await** b" to be $\{p\}$ — we can also add any program point that follows p until we reach an **await** or **return** instruction, because after p is executed the processor will not be released until the next **await** or **return** instruction.

In our implementation we support the condition "$x \mathrel{!=} \textbf{null}$", and numerical conditions that involve one field and one number, e.g., "$x \mathrel{!=} N$", "$x == N$" or "$x > N$" where N is an integer. For "$x \mathrel{!=} \textbf{null}$" we seek an instruction that assign x a non-**null** reference, and for the above numerical conditions we seek an instruction that assign x a value that is different, equal or greater than N. Moreover, we require that those are the only assignments to x in the program.

Example 8. Consider the program of Fig. 2. Applying the above technique to the **await** instruction at program point 32, we can prove that the only instruction that makes the corresponding condition satisfiable is the one at program point 17. Moreover, there is no instruction that makes the condition unsatisfiable. Thus, we add program point 17 to mhhseed(33).

4.4 Under-Approximating the Set $\mathcal{U}$ of Unique Program Points

A sufficient condition for a program point p to be unique is that it belongs to a method m that is unique (i.e., m is executed only once) and that it is not part of a loop. The challenge is how to prove that a given method m is unique.

Recall that in the semantics of Sect. 2.2 tasks never disappear once finished, but they rather remain at their exit point during the remaining execution. This behavior is correctly handled by the basic MHP analysis. Consequently, if a method m executes more than once, eventually some of its program points will "run" in parallel. This is true even if these instances of m execute one after the completion of the other. Therefore, if there are no program points p_1 and p_2 in m such that $(p_1, p_2) \in \tilde{\mathcal{E}}_P$ (the result of the basic MHP analysis that we have assumed to be available), we can safely conclude that m is unique.

Example 9. We have seen in Example 6 that $17 \in \tilde{E}_{35}$, i.e., program point 17 is in the MHH-set of program point 35. Since 17 does not appear in a loop and method init is unique, we can conclude that 17 is in the MHF-set of 35.

4.5 Further Improvements

Eliminating MHP pairs using the unique methods structure. We can extend the computed MHF sets by analyzing the structure of unique methods. In the example of Fig. 2, we cannot conclude that $15 \in \mathtt{mhfset}(35)$ since 15 is not unique. Moreover, $15 \notin \mathtt{mhhset}(35)$ because the body of the loop might not be executed at all. Fortunately, 17 still belongs to $\mathtt{mhfset}(35)$ and a syntactic analysis of method init allows us to conclude that $15 \in \mathtt{mhfset}(17)$. Given that 17 is guaranteed to execute before 35, we can conclude that $15 \in \mathtt{mhfset}(35)$.

Eliminating MHP pairs using the MHP graph $\mathcal{G}_P$. In the previous extension, we were able to discard all the program points of init (with respect to 35) but we cannot discard 22 because init is not unique. However, all instances of init are called from init and executed while init is in the loop, before 17. This fact is reflected in the MHP graph $\mathcal{G}_P$ (see Sect. 3). All the paths that allow us to reach 22 from 35 traverse one of the program points of init which are guaranteed to have finished by the time 35 is reached. In fact, 22 and 35 cannot happen in parallel. This can be detected by incorporating the MHF-sets in the corresponding MHP-graphs $\mathcal{G}_P$. Given a program point p its corresponding set of refined MHP pairs, denoted $\tilde{\mathcal{E}}_P^p$, is constructed as follows: (1) first remove all nodes that correspond to program points in $\mathtt{mhfset}(p)$ from the MHP graph $\mathcal{G}_P$; and (2) construct a set of MHP pairs from the modified graph using the same reachability conditions as in [4]. Then, given two program points p_1 and p_2, if $(p_1, p_2) \notin \tilde{\mathcal{E}}_P^{p_1} \cap \tilde{\mathcal{E}}_P^{p_2}$, the pair (p_1, p_2) can be safely eliminated from $\tilde{\mathcal{E}}_P$. With this improvement, we can guarantee that the method init will not be executed during the await 35. Therefore, we will be able to prove termination of the while loop in method go.

5 Experimental Evaluation

We have implemented our analysis in the SACO system. It can be used online through a web interface at http://costa.ls.fi.upm.es/saco by switching on the option "Condition synchronization MHP extension" in the "Settings-MHP" menu (it is done by default for the deadlock analysis in SACO). Experimental evaluation has been carried out using two industrial case studies: the Replication-System developed by Fredhopper® (1881 lines of code) and the TradingSystem (1187 lines) developed in the EU HATS project http://www.hats-project.eu. Their source code can be found at the SACO website.

For the ReplicationSystem (resp. TradingSystem) the total number of program point pairs is 90601 (resp. 18769). The naïve version of the analysis as described in Sect. 3.3 infers a total of 61278 (resp. 14829) MHP pairs. Out of these pairs, the enhanced analysis for condition synchronization is able to discard

698 (resp. 3128) pairs which correspond to infeasible pairs resulting from the naïve treatment of condition synchronization. The number of MHH-seeds that have been extracted from "await b" instructions is only 6 (resp. 1). These seeds contain the kind of information that is completely ignored in the naïve extension of the MHP analysis. Importantly, it can be seen that a small amount of seeds can result in a considerable amount of discarded pairs. It is specially notice-able the case of the TradingSystem where one MHH-seed allows us to discard 3128 pairs.

Note that the number of discarded pairs is quite small in comparison with the total number of pairs. However, the important conclusions are: (1) Many of the MHP pairs are trivially true and thus they are not relevant to prove later complex properties. For instance, it is common to have parts of a system that run in parallel independently from each other for performance purposes. We will obtain that their program points happen in parallel with each other. However, these do not have to be relevant when proving properties like termination or deadlock, since there is no interaction among these parts. (2) In the design of these case studies, futures variables have been preferred over condition synchronization but this will not be the case in general. (3) On the other hand, instructions of the form "await b" are often explicitly used to synchronize objects internally and therefore, it is likely that the discarded pairs with this technique are significant. (4) Finally, the false positives that have been discarded allow proving deadlock freedom of the ReplicationSystem (the previous analysis [10] could not prove it).

As regards efficiency, the analysis of the ReplicationSystem (resp. TradingSystem) has taken 9592 ms (resp. 3545 ms). Besides, although in the experiments we have computed the complete set of MHP pairs, for most applications, only a reduced set of points needs to be queried. Both the original MHP analysis and the MHF extension can be partially computed for specific program points. For example, the required time to compute the MHP analysis plus the MHF extension when queried by the mentioned deadlock analysis of the ReplicationSystem is less than 2 s. In conclusion, we argue that our experiments empirically show that our analysis is both accurate and efficient in practice.

6   Conclusions and Related Work

Static analysis of concurrent programs is considered complex and computation-ally expensive. The analysis is expensive because one needs to consider all possi-ble interleavings of processes. Cooperative scheduling, as used in our concurrency model, greatly alleviates this problem because interleavings can only occur at explicit points (in contrast to preemptive scheduling in which interleavings must be considered at every point). As regards complexity of the analysis, one of the main challenges is on understanding the synchronization of processes. In the context of cooperative scheduling, the problem was partially solved in [4], where synchronization based on future variables was accurately treated. This work extends such previous analysis to handle synchronization on shared variables effectively, thus enabling the analysis of programs that use both mechanisms.

The MHP analyses of [2,14] for X10 focus on synchronization idioms that do not allow conditional synchronizations. In Java, condition synchronization can be simulated using wait-notify. The analysis of [17] supports wait-notify. However, in this context the relation between the waiting and notifying threads is easier to identify (in our case this is done by the seeds). The analysis of [7] does not support wait-notify. The analysis of [16] does not support any synchronization idiom of Java, it just analyzes the thread structure of the program. Recently, the basic MHP analysis has been extended to handle inter-procedural synchronization [6] in which the future variables can be passed as parameters so that it is possible to synchronize the execution with the termination of the tasks created outside the method in which they are awaited. This extension can be used directly in our framework as well.

Acknowledgements. We would like to thank the reviewers for their comments that have helped improve the quality and the presentation of the paper. This work was funded partially by the EU project FP7-ICT-610582 ENVISAGE: Engineering Virtualized Services (http://www.envisage-project.eu), the Spanish MINECO project TIN2012-38137, and the CM project S2013/ICE-3006.

References

1. Ericsson AB. Erlang Efficiency Guide, 5.8., 5th edn., October 2011. http://www.erlang.org/doc/efficiency_guide/users_guide.html
2. Agarwal, S., Barik, R., Sarkar, V., Shyamasundar, R.K.: May-happen-in-parallel analysis of x10 programs. In: Yelick, K.A., Mellor-Crummey, J.M. (eds.) Proceedings of PPOPP 2007, pp. 183–193. ACM (2007)
3. Albert, E., Arenas, P., Correas, J., Genaim, S., Gómez-Zamalloa, M., Puebla, G., Román-Díez, G.: Object-sensitive cost analysis for concurrent objects. Softw. Test. Verif. Reliab. **25**(3), 218–271 (2015)
4. Albert, E., Flores-Montoya, A.E., Genaim, S.: Analysis of may-happen-in-parallel in concurrent objects. In: Giese, H., Rosu, G. (eds.) FMOODS/FORTE -2012. LNCS, vol. 7273, pp. 35–51. Springer, Heidelberg (2012). doi:10.1007/978-3-642-30793-5_3
5. Albert, E., Flores-Montoya, A., Genaim, S., Martin-Martin, E.: Termination and cost analysis of loops with concurrent interleavings. In: Hung, D., Ogawa, M. (eds.) ATVA 2013. LNCS, vol. 8172, pp. 349–364. Springer, Heidelberg (2013). doi:10.1007/978-3-319-02444-8_25
6. Albert, E., Genaim, S., Gordillo, P.: May-happen-in-parallel analysis for asynchronous programs with inter-procedural synchronization. In: Blazy, S., Jensen, T. (eds.) SAS 2015. LNCS, vol. 9291, pp. 72–89. Springer, Heidelberg (2015). doi:10.1007/978-3-662-48288-9_5
7. Barik, R.: Efficient computation of may-happen-in-parallel information for concurrent java programs. In: Ayguadé, E., Baumgartner, G., Ramanujam, J., Sadayappan, P. (eds.) LCPC 2005. LNCS, vol. 4339, pp. 152–169. Springer, Heidelberg (2006). doi:10.1007/978-3-540-69330-7_11
8. Boer, F.S., Clarke, D., Johnsen, E.B.: A complete guide to the future. In: Nicola, R. (ed.) ESOP 2007. LNCS, vol. 4421, pp. 316–330. Springer, Heidelberg (2007). doi:10.1007/978-3-540-71316-6_22

9. Emmi, M., Lal, A., Qadeer, S.: Asynchronous programs with prioritized task-buffers. In: Proceedings of the ACM SIGSOFT 20th International Symposium on the Foundations of Software Engineering, FSE 2012, pp. 48:1–48:11. ACM, New York (2012)

10. Flores-Montoya, A.E., Albert, E., Genaim, S.: May-happen-in-parallel based dead-lock analysis for concurrent objects. In: Beyer, D., Boreale, M. (eds.) FMOODS/-FORTE -2013. LNCS, vol. 7892, pp. 273–288. Springer, Heidelberg (2013). doi:10.1007/978-3-642-38592-6_19

11. Giachino, E., Grazia, C.A., Laneve, C., Lienhardt, M., Wong, P.: Deadlock Analysis of Concurrent Objects - Theory and Practice (2013)

12. Haller, P., Odersky, M.: Scala actors: unifying thread-based and event-based programming. Theor. Comput. Sci. **410**(2–3), 202–220 (2009)

13. Johnsen, E.B., Hähnle, R., Schäfer, J., Schlatte, R., Steffen, M.: ABS: a core language for abstract behavioral specification. In: Aichernig, B.K., Boer, F.S., Bonsangue, M.M. (eds.) FMCO 2010. LNCS, vol. 6957, pp. 142–164. Springer, Heidelberg (2011). doi:10.1007/978-3-642-25271-6_8

14. Lee, J.K., Palsberg, J.: Featherweight X10: a core calculus for async-finish parallelism. In: Proceedings of PPoPP 2010, pp. 25–36. ACM (2010)

15. Milanova, A., Rountev, A., Ryder, B.G.: Parameterized object sensitivity for points-to analysis for Java. ACM Trans. Softw. Eng. Methodol. **14**, 1–41 (2005)

16. Naik, M., Park, C.-S., Sen, K., Gay, D.: Effective static deadlock detection. In: Proceedings of the 31st International Conference on Software Engineering, ICSE 2009, pp. 386–396. IEEE Computer Society, Washington, DC (2009)

17. Naumovich, G., Avrunin, G.S., Clarke, L.A.: An efficient algorithm for computing MHP information for concurrent Java programs. SIGSOFT Softw. Eng. Not. **24**(6), 338–354 (1999). 319252

18. Nielson, F., Nielson, H.R., Hankin, C.: Principles of Program Analysis, 2nd edn. Springer, Heidelberg (2005)

Using Dependent Types to Define Energy Augmented Semantics of Programs

Bernard van Gastel[1,2]([✉]), Rody Kersten[3], and Marko van Eekelen[1,2]

[1] Institute for Computing and Information Sciences,
Radboud University, Nijmegen, The Netherlands
{b.vangastel,m.vaneekelen}@cs.ru.nl
[2] Faculty of Management, Science and Technology,
Open University of the Netherlands, Heerlen, The Netherlands
{bernard.vangastel,marko.vaneekelen}@ou.nl
[3] Carnegie Mellon Silicon Valley, Moffett Field, CA, USA
rody.kersten@sv.cmu.edu

Abstract. Energy is becoming a key resource for IT systems. Hence, it can be essential for the success of a system under development to be able to derive and optimise its resource consumption. For large IT systems, compositionality is a key property in order to be applicable in practice. If such a method is hardware parametric, the effect of using different algorithms or running the same software on different hardware configurations can be studied. This article presents a hardware-parametric, compositional and precise type system to derive energy consumption functions. These energy functions describe the energy consumption behaviour of hardware controlled by the software. This type system has the potential to predict energy consumptions of algorithms and hardware configurations, which can be used on design level or for optimisation.

1 Introduction

Green computing is an important emerging field within computer science. Much attention is devoted to developing energy-efficient systems, with a traditional focus on hardware. However, this hardware is controlled by software, which therefore has an energy-footprint as well.

A few options are available to a programmer who wants to write energy-efficient code. First, the programmer can look for programming guidelines and design patterns, which generally produce more energy-efficient programs, e.g. [1]. Then, he/she might make use of a compiler that optimizes for energy-efficiency, e.g. [2]. If the programmer is lucky, there is an energy analysis available for his specific platform, such as [3] for a processor that is modeled in SIMPLESCALAR (note that this only analyses the energy consumption of the processor, no other components). However, for most platforms this is not a viable option. In that case, the programmer might use dynamic analysis with a measurement set-up. This, however, is not a trivial task and requires a complex set-up [4]. Moreover, it only yields information for a specific test case.

© Springer International Publishing Switzerland 2016
M. van Eekelen and U. Dal Lago (Eds.): FOPARA 2015, LNCS 9964, pp. 20–39, 2016.
DOI: 10.1007/978-3-319-46559-3_2

Our Approach. We propose a dependent type system, which can be used to analyse energy consumption of software, with respect to a set of hardware component models. Such models can be constructed once, using a measurement set-up. Alternatively, they might be derived from hardware specifications. This type system is precise, in the sense that no over-approximation is used. By expressing energy analysis as a dependent type system, one can easily reuse *energy signatures* for functions which were derived earlier. This makes this new dependent type system modular and composable. Furthermore, the use of energy signatures that form a precise description of energy consumption can be a flexible, modular basis for various kinds of analyses and estimations using different techniques (e.g. lower or upper bound static analysis using approximations or average case analysis using dynamic profiling information).

The presented work is related to our results in [5], where we present an over-approximating energy analysis for software, parametric with respect to a hardware model. That analysis is based on an energy-aware Hoare logic and operates on a simple while-language. It is implemented in the tool ECALOGIC [6]. This previous work poses many limitations on hardware models in order to over-approximate and requires re-analysis of the body of a function *at each function call*.

The most important contributions of this article are:

1. A *dependent type system* that, for the analysed code, captures both energy consumption and the effect on the state of hardware components.
2. Through the use of energy type signatures the system is *compositional*, making it more suitable for larger IT systems.
3. The dependent type system derives *precise information*. This in contrast to the energy aware Hoare logic in [5], which uses over-approximations for conditionals and loops.
4. Compared with [5] many restrictions on component models and component functions are now removed. Effectively all that is required now, is that the power consumption of a device can be modelled by a finite state machine in which states correspond to level of power draw and state transitions correspond to changes of power draw level. State transitions occur due to (external) calls of component functions. The state transition itself may consume a certain amount of energy. This makes the system very suited for control software for various devices where the actuators are performed directly without the need for synchronisation.
5. The dependently typed energy signatures can form a solid, modular basis for various approximations with various different static or dynamic techniques.
6. The artificial environment ρ that was introduced in [5] to incorporate user verified properties, is not needed in this dependent type setting. Using dependent types, such properties can be incorporated in a more *elegant* way.

We start the paper with a discussion of the considered language and its (energy) semantics in Sect. 2. Next, we need a type system to express every variable in terms of input variables in the appropriate scope (block, function or program). We introduce a basic dependent type system in Sect. 3 specifically for this problem. Continuing in Sect. 4, we introduce the main type system which derives from each statement and expression both an energy bound and a component state effect. To demonstrate the type system we analyse and compare two example programs in Sect. 5. We conclude this article with a discussion, describing future work, and a conclusion.

2   Hybrid Modelling: Language and Semantics

Modern electronic systems typically consist of both hardware and software. As we aim to analyse energy consumption of such *hybrid* systems, we consider hardware and software in a single modelling framework. Software plays a central role, controlling various hardware components. Our analysis works on a simple, special purpose language, called ECA. This language has a special construct for calling functions on hardware components. It is assumed that models exist for all the used hardware components, which model the energy consumption characteristics of these components, as well as the state changes induced by and return values of component function calls.

The ECA language is described in Sect. 2.1. Modelling of hardware components is discussed in Sect. 2.2. Energy-aware semantics for ECA are discussed in Sect. 2.3.

2.1   ECA Language

The grammar for the ECA language is defined below. We presume there is a way to differentiate between identifiers that represents variables (VAR), function names (FUNNAME), components (COMPONENT), and constants (CONST).

$\langle$FUNDEF$\rangle$ ::= 'function' $\langle$FUNNAME$\rangle$ '(' $\langle$VAR$\rangle$ ')' 'begin' $\langle$EXPR$\rangle$ 'end'

$\langle$BIN-OP$\rangle$ ::= '+' | '-' | '*' | '>' | '>=' | '==' | '!=' | '<=' | '<' | 'and' | 'or'

$\langle$EXPR$\rangle$ ::= $\langle$CONST$\rangle$ | '-' $\langle$CONST$\rangle$ | $\langle$INPUT$\rangle$ | $\langle$VAR$\rangle$
          | $\langle$VAR$\rangle$ ':=' $\langle$EXPR$\rangle$ | $\langle$EXPR$\rangle$ $\langle$BIN-OP$\rangle$ $\langle$EXPR$\rangle$
          | $\langle$COMPONENT$\rangle$ '.' $\langle$FUNNAME$\rangle$ '(' $\langle$EXPR$\rangle$ ')'
          | $\langle$FUNNAME$\rangle$ '(' $\langle$EXPR$\rangle$ ')'
          | $\langle$STMT$\rangle$ ',' $\langle$EXPR$\rangle$

$\langle$STMT$\rangle$ ::= 'skip' | $\langle$STMT$\rangle$ ';' $\langle$STMT$\rangle$ | $\langle$EXPR$\rangle$
          | 'if' $\langle$EXPR$\rangle$ 'then' $\langle$STMT$\rangle$ 'else' $\langle$STMT$\rangle$ 'end'
          | 'repeat' $\langle$EXPR$\rangle$ 'begin' $\langle$STMT$\rangle$ 'end'
          | $\langle$FUNDEF$\rangle$ $\langle$STMT$\rangle$

The only supported type in the ECA language is a signed integer. There are no explicit booleans. The value 0 is handled as *false*, any other value is handled as *true*. The absence of global variables and the by-value passing of variables to functions imply that functions do not have side-effects on the program state. Functions are statically scoped. Recursion is not currently supported.

The language has an explicit construct for operations on hardware components (e.g. memory, storage or network devices). This allows us to reason about components in a straight-forward manner. Functions on components have a single parameter and always return a value. The notation $C.f$ refers to a function f of a component C.

The language supports a repeat construct, which makes the bound an obvious part of the loop and removes the need for evaluating the loop guard with every iteration (as with the while construct). The loop bound is evaluated once before executing the loop bodies. The repeat construct is used for presentation purposes, without loss of generality, since the more commonly used while construct has the same expressive power, but is more complex to present.

2.2   Hardware Component Modelling

In order to reason about hybrid systems, we need to model hardware components. A component model consists of a *component state* and a set of *component functions* which can change the component state. A component model must capture the behaviour of the hardware component with respect to energy consumption. Component models are used to derive dependent types signifying the energy consumption of the modelled hybrid system. The model can be based on measurements or detailed hardware specifications. Alternatively, a generic component model might be used (e.g. for a generic hard disk drive).

A component state $C.s$ is a collection of variables of any type. They can signify e.g. that the component is on, off or in stand-by. A component function is modelled by two functions: one that produces the return value (rv_f) and one that updates the (component) state (δ_f). A component can have multiple component functions. Any state change in components can only occur during a component function call and therefore is made explicit in the source code.

Each component has a power draw, which depends on the component state. The function ϕ in the component model maps the component state to a power draw. The result of this function is used to calculate *time-dependent energy consumption* in the td_s function. Function td_s gets the elapsed time and the set of component models as input and results in a function that updates the component states.

2.3   Energy-Aware Semantics

We use a fairly standard semantics for our ECA language, to which we add energy-awareness. We do not list the basic semantics here. The energy-aware semantics are given in Fig. 1. The interested reader might refer to [7] for a listing of the full non-energy-aware semantics of a previous version of ECA.

Components consume energy in two distinct ways. A component function might directly induce a *time-independent* energy consumption. Apart from that, it can change the state of the component, affecting its *time-dependent* energy consumption. To be able to calculate time-dependent consumption, we need a

$$\frac{}{\Delta^s \vdash \langle i,\, \sigma,\, \Gamma \rangle \xrightarrow{e} \langle \mathcal{I}(i),\, \sigma,\, td_s(\Gamma, t_{\text{input}}) \rangle} \text{ (esInput)} \qquad \frac{}{\Delta^s \vdash \langle c,\, \sigma,\, \Gamma \rangle \xrightarrow{e} \langle \mathcal{Z}(c),\, \sigma,\, td_s(\Gamma, t_{\text{const}}) \rangle} \text{ (esConst)}$$

$$\frac{}{\Delta^s \vdash \langle x,\, \sigma,\, \Gamma \rangle \xrightarrow{e} \langle \sigma(x),\, \sigma,\, td_s(\Gamma, t_{\text{var}}) \rangle} \text{ (esVar)}$$

$$\frac{\Delta^s \vdash \langle e_1,\, \sigma,\, \Gamma \rangle \xrightarrow{e} \langle n,\, \sigma',\, \Gamma' \rangle \qquad \Delta^s \vdash \langle e_2,\, \sigma',\, \Gamma' \rangle \xrightarrow{e} \langle m,\, \sigma'',\, \Gamma'' \rangle \qquad n \mathbin{\square} m = p}{\Delta^s \vdash \langle e_1 \mathbin{\square} e_2,\, \sigma,\, \Gamma \rangle \xrightarrow{e} \langle p,\, \sigma'',\, td_s(\Gamma'', t_{\square}) \rangle} \text{ (esBinOp)}$$

$$\frac{\Delta^s \vdash \langle e_1,\, \sigma,\, \Gamma \rangle \xrightarrow{e} \langle n,\, \sigma',\, \Gamma' \rangle}{\Delta^s \vdash \langle x := e_1,\, \sigma,\, \Gamma \rangle \xrightarrow{e} \langle n,\, \sigma'[x \leftarrow n],\, td_s(\Gamma', t_{\text{assign}}) \rangle} \text{ (esAssign)}$$

$$\frac{\Delta^s \vdash \langle e_1,\, \sigma,\, \Gamma \rangle \xrightarrow{e} \langle a,\, \sigma',\, \Gamma' \rangle \qquad \Gamma'' = td_s(\Gamma', t)}{\Delta^s, C.f = (t, \delta, rv) \vdash \langle C.f(e_1),\, \sigma,\, \Gamma \rangle \xrightarrow{e} \langle rv(\Gamma''(C), a),\, \sigma',\, \Gamma''[C \leftarrow \delta(\Gamma''(C), a)] \rangle} \text{ (esCallCmpF)}$$

$$\frac{\Delta^s \vdash \langle e_1,\, \sigma,\, \Gamma \rangle \xrightarrow{e} \langle a,\, \sigma',\, \Gamma' \rangle \qquad \Delta^s \vdash \langle e_f,\, [x \mapsto a],\, \Gamma' \rangle \xrightarrow{e} \langle n,\, \sigma'',\, \Gamma'' \rangle}{\Delta^s, f = (e_f, x) \vdash \langle f(e_1),\, \sigma,\, \Gamma \rangle \xrightarrow{e} \langle n,\, \sigma',\, \Gamma'' \rangle} \text{ (esCallF)}$$

$$\frac{\Delta^s \vdash \langle S_1,\, \sigma,\, \Gamma \rangle \xrightarrow{s} \langle \sigma',\, \Gamma' \rangle \qquad \Delta^s \vdash \langle e_1,\, \sigma',\, \Gamma' \rangle \xrightarrow{e} \langle n,\, \sigma'',\, \Gamma'' \rangle}{\Delta^s \vdash \langle S_1, e_1,\, \sigma,\, \Gamma \rangle \xrightarrow{e} \langle n,\, \sigma'',\, \Gamma'' \rangle} \text{ (esExprConcat)}$$

$$\frac{\Delta^s \vdash \langle e_1,\, \sigma,\, \Gamma \rangle \xrightarrow{e} \langle n,\, \sigma',\, \Gamma' \rangle}{\Delta^s \vdash \langle e_1,\, \sigma,\, \Gamma \rangle \xrightarrow{s} \langle \sigma',\, \Gamma' \rangle} \text{ (esExprAsStmt)} \qquad \frac{}{\Delta^s \vdash \langle \text{skip},\, \sigma,\, \Gamma \rangle \xrightarrow{s} \langle \sigma,\, \Gamma \rangle} \text{ (esSkip)}$$

$$\frac{\Delta^s \vdash \langle S_1,\, \sigma,\, \Gamma \rangle \xrightarrow{s} \langle \sigma',\, \Gamma' \rangle \qquad \Delta^s \vdash \langle S_2,\, \sigma',\, \Gamma' \rangle \xrightarrow{s} \langle \sigma'',\, \Gamma'' \rangle}{\Delta^s \vdash \langle S_1; S_2,\, \sigma,\, \Gamma \rangle \xrightarrow{s} \langle \sigma'',\, \Gamma'' \rangle} \text{ (esStmtConcat)}$$

$$\frac{\Delta^s \vdash \langle e_1,\, \sigma,\, \Gamma \rangle \xrightarrow{e} \langle 0,\, \sigma',\, \Gamma' \rangle \qquad \Delta^s \vdash \langle S_2,\, \sigma',\, td_s(\Gamma', t_{\text{if}}) \rangle \xrightarrow{s} \langle \sigma'',\, \Gamma'' \rangle}{\Delta^s \vdash \langle \text{if } e_1 \text{ then } S_1 \text{ else } S_2 \text{ end},\, \sigma,\, \Gamma \rangle \xrightarrow{s} \langle \sigma'',\, \Gamma'' \rangle} \text{ (esIf-False)}$$

$$\frac{\Delta^s \vdash \langle e_1,\, \sigma,\, \Gamma \rangle \xrightarrow{e} \langle n,\, \sigma',\, \Gamma' \rangle \qquad n \neq 0 \qquad \Delta^s \vdash \langle S_1,\, \sigma',\, td_s(\Gamma', t_{\text{if}}) \rangle \xrightarrow{s} \langle \sigma'',\, \Gamma'' \rangle}{\Delta^s \vdash \langle \text{if } e_1 \text{ then } S_1 \text{ else } S_2 \text{ end},\, \sigma,\, \Gamma \rangle \xrightarrow{s} \langle \sigma'',\, \Gamma'' \rangle} \text{ (esIf-True)}$$

$$\frac{n \leq 0}{\Delta^s \vdash \langle (S_1, n),\, \sigma,\, \Gamma \rangle \xrightarrow{s} \langle \sigma,\, \Gamma \rangle} \text{ (esRepeatBase)}$$

$$\frac{n > 0 \qquad \Delta^s \vdash \langle S_1,\, \sigma,\, \Gamma \rangle \xrightarrow{s} \langle \sigma',\, \Gamma' \rangle \qquad \Delta^s \vdash \langle (S_1, n-1),\, \sigma',\, \Gamma' \rangle \xrightarrow{s} \langle \sigma'',\, \Gamma'' \rangle}{\Delta^s \vdash \langle (S_1, n),\, \sigma,\, \Gamma \rangle \xrightarrow{s} \langle \sigma'',\, \Gamma'' \rangle} \text{ (esRepeatLoop)}$$

$$\frac{\Delta^s \vdash \langle e_1,\, \sigma,\, \Gamma \rangle \xrightarrow{e} \langle n,\, \sigma',\, \Gamma' \rangle \qquad \Delta^s \vdash \langle (S_1, n),\, \sigma',\, \Gamma' \rangle \xrightarrow{s} \langle \sigma'',\, \Gamma'' \rangle}{\Delta^s \vdash \langle \text{repeat } e_1 \text{ begin } S_1 \text{ end},\, \sigma,\, \Gamma \rangle \xrightarrow{s} \langle \sigma'',\, \Gamma'' \rangle} \text{ (esRepeat)}$$

$$\frac{\Delta^s, f = (e_1, x) \vdash \langle S_1,\, \sigma,\, \Gamma \rangle \xrightarrow{s} \langle \sigma',\, \Gamma' \rangle}{\Delta^s \vdash \langle \text{function } f(x) \text{ begin } e_1 \text{ end } S_1,\, \sigma,\, \Gamma \rangle \xrightarrow{s} \langle \sigma',\, \Gamma' \rangle} \text{ (esFuncDef)}$$

Fig. 1. Energy-aware semantics. Note that i stands for an INPUT term, c for a CONST term, x for a VAR term, e for a EXPR term, S for a STMT term, $\square$ for a BIN-OP, f for a FUNNAME, and C for a COMPONENT.

basic timing analysis. Each construct in the language therefore has an associated execution time, for instance t_{if}. Component functions have a constant time consumption that is part of its function signature, along with δ and rv, as described in Sect. 2.2.

The function environment Δ^s (s for semantics) contains both the aforementioned component function signatures (triples of duration t, state transition function δ and return value function rv) and function definitions (pairs of function body e and parameter x). Component models are collected in component state environment Γ. Time-dependent energy consumption is accounted for separately in Γ for each component by the td_s function, which gets Γ and a time period t as input and results in a new component state environment. Time independent energy usage can be accounted for by also assigning a constant energy cost to component state update function δ.

We differentiate two kinds of reductions. Expressions reduce from a triple of the expression, program state σ and component state environment Γ, to a triple of a value, a new program state and a new component state environment, with the $\xrightarrow{e}$ operator. As statements do not result in a value, they reduce from a triple of the statement, σ, and Γ to a pair of a new program state and a new component state environment, with the $\xrightarrow{s}$ operator.

We use the following notation for substitution: $\sigma[x \leftarrow a]$. With $[x \mapsto a]$, we construct a new environment in which x has the value a.

There are three rules for the repeat loop. The one labelled *esRepeat* calculates the value of expression e, i.e. the number of iterations for the loop. The *esRepeatLoop* rule then handles each iteration, until the number of remaining iterations is 0. At that point, the evaluation of the loop is ended with the *esRepeatBase* rule. We use a tuple with as first argument the statement to be evaluated and as second the number of times the statement should be evaluated to differentiate it from normal evaluation rules.

3   Basic Dependent Type System

Before we can introduce a dependent type system that can be used to derive energy consumption expressions of an ECA program, we need to define a standard dependent type system to reason about variable values. We separately introduce these dependent type systems for presentation purposes. We describe the dependent type system deriving types signifying energy consumption in Sect. 4.

Note that, instead of direct expressions over inputs, we derive functions that calculate a value based on values of the input. This allows us to combine these functions using function composition and to reuse them as signatures for methods and parts of the code.

We will start with a series of definitions. A program state environment called *PState* is a function from a variable identifier to a value, i.e. *PState* : *Var* $\mapsto$ *Value*. Values are of type $\mathbb{Z}$, like the data type in ECA. A component state *CState* is collection of states of components that the component function can work on. A *value function* is a function that, given a (program and component) state, will yield a concrete value, i.e. its signature is *PState* $\times$ *CState* $\rightarrow$ *Value*. A *state update function* is a function from *PState* $\times$ *CState* to *PState* $\times$ *CState*. Such a function expresses changes to the state, caused by the execution of statements or expressions.

$$\frac{}{\Delta^v \vdash i : \langle Lookup_i,\ id \rangle}\ (\text{btInput}) \qquad \frac{}{\Delta^v \vdash c : \langle Const_{\mathcal{N}(c)},\ id \rangle}\ (\text{btConst})$$

$$\frac{}{\Delta^v \vdash x : \langle Lookup_x,\ id \rangle}\ (\text{btVar})$$

$$\frac{\Delta^v \vdash e_1 : \langle V_1,\ \Sigma_1 \rangle \qquad \Delta^v \vdash e_2 : \langle V_2,\ \Sigma_2 \rangle \qquad \boxdot \in \text{Bin-Op}}{\Delta^v, \Sigma \vdash e_1 \boxdot e_2 : \langle V_1 \boxdot (\Sigma_1 \ggg V_2),\ \Sigma_1 \ggg \Sigma_2 \rangle}\ (\text{btBinOp})$$

$$\frac{\Delta^v \vdash e : \langle V,\ \Sigma \rangle}{\Delta^v \vdash x{:=}e : \langle V,\ \Sigma \ggg Assign_x(V) \rangle}\ (\text{btAssign})$$

$$\frac{\Delta^v \vdash e : \langle V_{ex},\ \Sigma_{ex} \rangle}{\Delta^v,\ C.f = (x_{f'}, V_{f'}, \Sigma_{f'}) \vdash \\ C.f(e) : \langle \overline{[x_{f'} \leftarrow V_{ex}, \Sigma_{ex}]} \ggg V_{f'},\ Split(\Sigma_{ex}, \overline{[x_{f'} \leftarrow V_{ex}, \Sigma_{ex}]} \ggg \Sigma_{f'}) \rangle}\ (\text{btCallCmpF})$$

$$\frac{\Delta^v \vdash e : \langle V_{ex},\ \Sigma_{ex} \rangle}{\Delta^v,\ f = (x_{f'}, V_{f'}, \Sigma_{f'}) \vdash \\ f(e) : \langle \overline{[x_{f'} \leftarrow V_{ex}, \Sigma_{ex}]} \ggg V_{f'},\ Split(\Sigma_{ex}, \overline{[x_{f'} \leftarrow V_{ex}, \Sigma_{ex}]} \ggg \Sigma_{f'}) \rangle}\ (\text{btCallF})$$

$$\frac{\Delta^v \vdash S : \Sigma_{st} \qquad \Delta^v \vdash e : \langle V_{ex},\ \Sigma_{ex} \rangle}{\Delta^v \vdash S, e : \langle \Sigma_{st} \ggg V_{ex},\ \Sigma_{st} \ggg \Sigma_{ex} \rangle}\ (\text{btExprConcat})$$

$$\frac{\Delta^v \vdash e : \langle V,\ \Sigma \rangle}{\Delta^v \vdash e : \Sigma}\ (\text{btExprAsStmt}) \qquad \frac{}{\Delta^v \vdash \text{skip} : id}\ (\text{btSkip})$$

$$\frac{\Delta^v \vdash S_1 : \Sigma_1 \qquad \Delta^v \vdash S_2 : \Sigma_2}{\Delta^v \vdash S_1; S_2 : \Sigma_1 \ggg \Sigma_2}\ (\text{btStmtConcat})$$

$$\frac{\Delta^v \vdash e : \langle V_{ex},\ \Sigma_{ex} \rangle \qquad \Delta^v \vdash S_t : \Sigma_t \qquad \Delta^v \vdash S_f : \Sigma_f}{\Delta^v \vdash \text{if } e \text{ then } S_t \text{ else } S_f \text{ end} : if(V_{ex}, \Sigma_{ex} \ggg \Sigma_t, \Sigma_{ex} \ggg \Sigma_f)}\ (\text{btIf})$$

$$\frac{\Delta^v \vdash e : \langle V_{ex},\ \Sigma_{ex} \rangle \qquad \Delta^v \vdash S : \Sigma_{st}}{\Delta^v \vdash \text{repeat } e \text{ begin } S \text{ end} : repeat^v(V_{ex}, \Sigma_{ex}, \Sigma_{st})}\ (\text{btRepeat})$$

$$\frac{\Delta^v \vdash e : \langle V_{ex},\ \Sigma_{ex} \rangle \qquad \Delta^v, f = (x, V_{ex}, \Sigma_{ex}) \vdash S : \Sigma_{st}}{\Delta^v \vdash \text{function } f(x) \text{ begin } e \text{ end } S : \Sigma_{st}}\ (\text{btFuncDef})$$

Fig. 2. Basic dependent type system. For expressions it yields a tuple of a *value function* and a *state update function*. For statements the type system only derives a *state update function*.

To make the typing rules more clear, we explain a number of rules in more detail. The full typing rules can be found in Fig. 2. We will start with the variable access rule:

$$\frac{}{\Delta^v \vdash x : \langle Lookup_x,\ id \rangle}\ (\text{btVar})$$

All rules for evaluation of an expression return a tuple of a function returning the value of the expression when evaluated and a function that modifies the program state and component state. The former one captures the value, i.e. it is a state value function. The latter one captures the effect, i.e. it is a state update function. To access a variable, we return the *Lookup* function (defined below), which is parametrised by the variable that is returned. Variable access does not affect the state, hence for that part the identity function *id* is returned.

The *Lookup* function that the *btVar* rule depends on is defined as follows:

$$Lookup_x : PState \times CState \to Value$$
$$Lookup_x(ps, cs) = ps(x)$$

Likewise we need to define a function for the *btConst* rule, dealing with constants:

$$Const_x : PState \times CState \to Value$$
$$Const_x(ps, cs) = x$$

Before we can continue we need to introduce the $\ggg$ operator. We can compose state update functions with the $\ggg$ operator. Note that the composition is reversed with respect to standard mathematical function composition, in order to maintain the order of program execution. For now, T is $PState \times CState$. The $\ggg$ operator can be interpreted as: first apply the effect of the left operand, then execute the right operand on the resulting state. The $\ggg$ operator is defined as:

$$\ggg: (PState \times CState \to PState \times CState) \times (PState \times CState \to T)$$
$$\to (PState \times CState \to T)$$
$$(A \ggg B)(ps, cs) = B(ps', cs') \text{where}(ps', cs') = A(ps, cs)$$

Moving on, we can explain the assignment rule below, which assigns a value expressed by expression e to variable x. As an assignment does not modify the value of the expression, the part capturing the value is propagated from the rule deriving the type of x. More interesting is the effect that captures the state change when evaluating the rule. This consists of first applying the state change Σ^v of evaluating the expression e and thereafter replacing the variable x with the result of e (as done by the *Assign* function, defined below). This effect is reached with the $\ggg$ operator.

$$\frac{\Delta^v \vdash e : \langle V, \Sigma^v \rangle}{\Delta^v \vdash x{:=}e : \langle V, \Sigma^v \ggg Assign_x(V) \rangle} \text{ (btAssign)}$$

The operator for assigning a new value to a program variable in the type environment:

$$Assign_x : (PState \times CState \to Value) \to (PState \times CState \to PState \times CState)$$
$$Assign_x(V)(ps, cs) = (ps[x \mapsto V(ps, cs)], cs)$$

In order to support binary operations we need to define a higher order operator $\overline{\boxdot}$ where $\boxdot \in +, -, \times, \div, \ldots$. It evaluates the two arguments and combines the results using $\boxdot$. For now, T is $Value$.

$$\overline{\boxdot} : (PState \times CState \to T) \times (PState \times CState \to T)$$
$$\to (PState \times CState \to T)$$
$$(A \overline{\boxdot} B)(ps, cs) = A(ps, cs) \boxdot B(ps, cs)$$

The binary operation rule can now be introduced. The pattern used in the rule reoccurs multiple times if two expressions (or statements) need to be combined. First the value function representing the first expression, V_1, is evaluated and is combined with the result representing the second expression, V_2, e.g. using the

$\boxdot$ operator. But this second value function needs to be evaluated in the right context (state): evaluating the first expression can have side-effects and therefore change the program and component state, and influence the outcome of a value function (as the expression can include function calls, assignments, component function calls, etc.). To evaluate V_2 in the right context we first must apply the side effects of the first expressions using a state update function Σ_1, expressed as $\Sigma_1 \ggg V_2$. We can express the value function that represents the combination of two expressions as $V_1 \boxdot (\Sigma_1 \ggg V_2)$. The binary operation rule is defined as:

$$\frac{\Delta^v \vdash e_1 : \langle V_1, \Sigma_1 \rangle \qquad \Delta^v \vdash e_2 : \langle V_2, \Sigma_2 \rangle \qquad \boxdot \in \text{BIN-OP}}{\Delta^v, \Sigma \vdash e_1 \boxdot e_2 : \langle V_1 \boxdot (\Sigma_1 \ggg V_2), \Sigma_1 \ggg \Sigma_2 \rangle} \text{(btBinOp)}$$

Without explaining the rule in detail, we introduce the conditional operator. The *if* operator captures the behaviour of a conditional inside the dependent type. The first argument denotes a function expressing the value of the conditional. For now, T stands for a tuple $PState \times CState$.

$$if : (PState \times CState \to Value) \times (PState \times CState \to T)$$
$$\times (PState \times CState \to T)$$
$$\to (PState \times CState \to T)$$

$$if(c, then, else)(ps, cs) = \begin{cases} then(ps, cs) & \text{if} c(ps, cs) \neq 0 \\ else(ps, cs) & \text{if} c(ps, cs) = 0 \end{cases}$$

The *btRepeat* rule can be used to illustrate a more complex rule. The effect of repeat is the composition of evaluating the bound and evaluating the body a number of times. The latter is captured in a *repeat* function that is defined below. The resulting effect can be defined as follows:

$$\frac{\Delta^v \vdash e : \langle V_{ex}, \Sigma_{ex}^v \rangle \qquad \Delta^v \vdash S : \Sigma_{st}^v}{\Delta^v \vdash \textsf{repeat } e \textsf{ begin } S \textsf{ end} : repeat^v(V_{ex}, \Sigma_{ex}^v, \Sigma_{st}^v)} \text{(btRepeat)}$$

The *repeat* operator needed in the *btRepeat* rule captures the behaviour of a loop inside the dependent type. It gets a function that calculates the number of iterations from a type environment, as well as an environment update function for the loop body, and results in an environment update function that represents the effects of the entire loop. Because the value of the bound must be evaluated in the context before the effect is evaluated, we need an extra function. The actual recursion is in the $repeat^v$ function, also defined below.

$$repeat^v : (PState \times CState \to Value) \times (PState \times CState \to PState \times CState)$$
$$\times (PState \times CState \to PState \times CState)$$
$$\to (PState \times CState \to PState \times CState)$$

$$repeat^v(c, start, body)(ps, cs) = repeat'^v(c(ps, cs), body, ps', cs')$$
$$\text{where}(ps', cs') = start(ps, cs)$$

$$repeat'^v : \mathbb{Z} \times (PState \times CState \to PState \times CState) \times PState \times CState$$
$$\to PState \times CState$$

$$repeat'^v(n, body, ps, cs) = \begin{cases} (ps, cs) & \text{if } n \leq 0 \\ repeat'^v(n-1, body, ps', cs') & \text{if } n > 0 \\ \text{where} \\ \qquad (ps', cs') = body(ps, cs) \end{cases}$$

In order to explain the component call rule we need an operator for higher order substitution. This operator creates a new program environment, but retains the component state. It can even update the component state given a Σ function, which is needed because this Σ needs to be evaluated using the original program state. The definition is as follows:

$$\overline{[x \leftarrow V, \Sigma]} : Var \times (PState \times CState \to Value)$$
$$\times (PState \times CState \to PState \times CState)$$
$$\to (PState \times CState \to PState \times CState)$$
$$\overline{[x \leftarrow V, \Sigma]}(ps, cs) = ([x \mapsto V(ps, cs)], cs')\text{where}(\_, cs') = \Sigma(ps, cs)$$

We also need an additional operator $Split$, because the program state is isolated but the component state is not. $Split$ forks the evaluation into two state update functions and joins the results together. The first argument defines the resulting program state, the second defines the resulting component state.

$$Split : (PState \times CState \to PState \times CState)$$
$$\times (PState \times CState \to PState \times CState)$$
$$\to (PState \times CState \to PState \times CState)$$
$$Split(A, B)(ps, cs) = (ps', cs') \quad \text{where}(ps', \_) = A(ps, cs)$$
$$\text{and}(\_, cs') = B(ps, cs)$$

The component functions and normal functions are stored in the environment Δ^v. For each function (both component and language) a triple is stored, which states the variable name of the argument, a value function that represents the return value, and a state update function that represents the effect on the state of executing the called function. We assume the component function calls are already in the environment, function definitions in the language can be placed in the environment by the function definition rule $btFuncDef$. The component function call can now be easily expressed. Likewise we can define the function call, but for brevity it is omitted in the text.

$$\frac{\Delta^v \vdash e : \langle V_{ex}, \Sigma_{ex} \rangle}{\begin{array}{l} \Delta^v, C.f = (x_{f'}, V_{f'}, \Sigma_{f'}) \vdash \\ \quad C.f(e) : \langle \overline{[x_{f'} \leftarrow V_{ex}, \Sigma_{ex}]} \ggg V_{f'}, \ Split(\Sigma_{ex}, \overline{[x_{f'} \leftarrow V_{ex}, \Sigma_{ex}]} \ggg \Sigma_{f'}) \rangle \end{array}} \text{(btCallCmpF)}$$

$$\frac{}{\Delta^{ec} \vdash i : \langle ..., ..., td^{ec}(t_{\text{input}}) \rangle} \text{ (etInput)} \qquad \frac{}{\Delta^{ec} \vdash c : \langle ..., ..., td^{ec}(t_{\text{const}}) \rangle} \text{ (etConst)}$$

$$\frac{}{\Delta^{ec} \vdash x : \langle ..., ..., td^{ec}(t_{\text{var}}) \rangle} \text{ (etVar)}$$

$$\frac{\Delta^{ec} \vdash e_1 : \langle ..., \Sigma_1, E_1 \rangle \qquad \Delta^{ec} \vdash e_2 : \langle ..., \Sigma_2, E_2 \rangle \qquad \boxdot \in \text{Bin-Op}}{\Delta^{ec} \vdash e_1 \boxdot e_2 : \langle ..., ..., E_1 \mp (\Sigma_1 \ggg (E_2 \mp (\Sigma_2 \ggg td^{ec}(t_\boxdot)))) \rangle} \text{ (etBinOp)}$$

$$\frac{\Delta^{ec} \vdash e : \langle ..., \Sigma, E \rangle}{\Delta^{ec} \vdash x{:=}e : \langle ..., ..., E \mp (\Sigma \ggg td^{ec}(t_{\text{assign}})) \rangle} \text{ (etAssign)}$$

$$\frac{\Delta^{ec} \vdash e : \langle V_{ex}, \Sigma_{ex}, E_{ex} \rangle}{\Delta^{ec}, C.f = (x_{f'}, ..., ..., E_{f'}, t_{f'}) \vdash \quad C.f(e) : \langle ..., ..., E_{ex} \mp ([x_{f'} \leftarrow V_{ex}, \Sigma_{ex}] \ggg (td^{ec}(t_{f'}) \mp E_{f'})) \rangle} \text{ (etCallCmpF)}$$

$$\frac{\Delta^{ec} \vdash e : \langle V_{ex}, \Sigma_{ex}, E_{ex} \rangle}{\Delta^{ec}, f = (x_{f'}, ..., ..., E_{f'}) \vdash f(e) : \langle ..., ..., E_{ex} \mp ([x_{f'} \leftarrow V_{ex}, \Sigma_{ex}] \ggg E_{f'}) \rangle} \text{ (etCallF)}$$

$$\frac{\Delta^{ec} \vdash S : \langle \Sigma_{st}, E_{st} \rangle \qquad \Delta^{ec} \vdash e : \langle ..., ..., E_{ex} \rangle}{\Delta^{ec} \vdash S, e : \langle ..., ..., E_{st} \mp (\Sigma_{st} \ggg E_{ex}) \rangle} \text{ (etExprConcat)}$$

$$\frac{\Delta^{ec} \vdash e : \langle ..., \Sigma, E \rangle}{\Delta^{ec} \vdash e : \langle \Sigma, E \rangle} \text{ (etExprAsStmt)} \qquad \frac{}{\Delta^{ec} \vdash \text{skip} : \langle ..., zero \rangle} \text{ (etSkip)}$$

$$\frac{\Delta^{ec} \vdash S_1 : \langle \Sigma_1, E_1 \rangle \qquad \Delta^{ec} \vdash S_2 : \langle ..., E_2 \rangle}{\Delta^{ec} \vdash S_1; S_2 : \langle ..., E_1 \mp (\Sigma_1 \ggg E_2) \rangle} \text{ (etStmtConcat)}$$

$$\frac{\Delta^{ec} \vdash e : \langle V_{ex}, \Sigma_{ex}, E_{ex} \rangle \qquad \Delta^{ec} \vdash S_t : \langle ..., E_t \rangle \qquad \Delta^{ec} \vdash S_f : \langle ..., E_f \rangle}{\Delta^{ec} \vdash \text{if } e \text{ then } S_t \text{ else } S_f \text{ end} : \langle ..., E_{ex} \mp (\Sigma_{ex} \ggg td^{ec}(t_{\text{if}})) \mp if(V_{ex}, \Sigma_{ex} \ggg E_t, \Sigma_{ex} \ggg E_f) \rangle} \text{ (etIf)}$$

$$\frac{\Delta^{ec} \vdash e : \langle V_{ex}, \Sigma_{ex}, E_{ex} \rangle \qquad \Delta^{ec} \vdash S : \langle \Sigma_{st}, E_{st} \rangle}{\Delta^{ec} \vdash \text{repeat } e \text{ begin } S \text{ end} : \langle ..., E_{ex} \mp repeat^{ec}(V_{ex}, \Sigma_{ex}, E_{st}, \Sigma_{st}) \rangle} \text{ (etRepeat)}$$

$$\frac{\Delta^{ec} \vdash e : \langle V_{ex}, \Sigma_{ex}, E_{ex} \rangle \qquad \Delta^{ec}, f = (x, V_{ex}, \Sigma_{ex}, E_{ex}) \vdash S : \langle \Sigma_{st}, E_{st} \rangle}{\Delta^{ec} \vdash \text{function } f(x) \text{ begin } e \text{ end } S : \langle E_{st}, \Sigma_{st} \rangle} \text{ (etFuncDef)}$$

Fig. 3. Energy-aware dependent type system. It adds an element to the resulting tuples of Fig. 2 signifying the energy consumption of executing the statement or expression.

4   Energy-Aware Dependent Type System

In this section, we present the complete dependent type system, which can be used to determine the energy consumption of ECA programs. The type system presented here should be viewed as an extension of the basic dependent type system presented in Sect. 3. Each time the three dot symbol $\cdots$ is used, that part of the rule is unchanged with respect to the previous section. The type system adds an element to the results of the previous section, signifying the energy bound. Expressions yield a triple, statements yield a pair. The added element is a function that, when evaluated on a concrete environment and component state, results in a concrete energy consumption or energy cost.

An important energy judgment is the td^{ec} function. To account for time-dependent energy usage, we need this function which calculates the energy usage of all the components over a certain time period. The td^{ec} function is a higher order function that takes the time it accounts for as argument. It results in a function that is just like the other function calculating energy bounds: it takes two arguments, a $PState$ and a $CState$. The definition is given below. It depends on the power draw function ϕ that maps a component state to an energy consumption (as explained in Sect. 2.2). The definition is as follows:

$$td^{ec} : Time \rightarrow (PState \times CState \rightarrow EnergyCost)$$
$$td^{ec}(t)(ps, cs) = \sum_{e \in cs} \phi(e) \times t$$

We can use this td^{ec} function to define a rule for variable lookup, as the only energy cost that is induced by variable access is the energy used by all components for the duration of the variable access. This is expressed in the $etVar$ rule, in which the dots correspond to Fig. 1 (the first dots are $Lookup_x$, the second id).

$$\frac{}{\Delta^{ec} \vdash x : \langle \dots, \dots, td^{ec}(t_{\mathrm{var}}) \rangle} \text{ (etVar)}$$

Using the $\overline{+}$ operator (introduced as $\overline{\Box}$, with type T equal to $EnergyCost$), we can define the energy costs of a binary operation. Although the rule looks complex, it uses pattern introduced in Sect. 3 for the binary operation. Basically, the energy cost of a binary operation is the cost of evaluating the operands, plus the cost of the binary operation itself. The binary operation itself only has an associated run-time and by this means induces energy consumption of components. This is expressed by the $td_{ec}(t_\Box)$ function. For the binary operation rule we need to apply this pattern twice, in a nested manner, as the time-dependent function must be evaluated in the context after evaluating both arguments. The (three) energy consumptions are added by $\overline{+}$. One can express the binary operation rule as:

$$\frac{\Delta^{ec} \vdash e_1 : \langle \dots, \Sigma_1, E_1 \rangle \qquad \Delta^{ec} \vdash e_2 : \langle \dots, \Sigma_2, E_2 \rangle}{\Delta^{ec} \vdash e_1 \Box e_2 : \langle \dots, \dots, E_1 \overline{+} (\Sigma_1 \ggg (E_2 \overline{+} (\Sigma_2 \ggg td^{ec}(t_\Box)))) \rangle} \text{ (etBinOp)}$$

Calculating the energy cost of the repeat loop can be calculated by evaluating an energy cost function for each loop iteration (in the right context). We therefore have to modify the $repeat^v$ rule to result in an energy cost, resulting in a new function definition of $repeat^{ec}$:

$$repeat^{ec} : (PState \times CState \rightarrow Value) \times (PState \times CState \rightarrow PState \times CState)$$
$$\times (PState \times CState \rightarrow EnergyCost)$$
$$\times (PState \times CState \rightarrow PState \times CState)$$
$$\rightarrow (PState \times CState \rightarrow EnergyCost)$$

$$repeat^{ec}(c, start, cost, body)(ps, cs) = repeat'^{ec}(c(ps, cs), cost, body, ps', cs')$$
$$\text{where}(ps', cs') = start(ps, cs)$$

$$repeat'^{ec} : \mathbb{Z} \times (PState \times CState \to EnergyCost)$$
$$\times (PState \times CState \to PState \times CState)$$
$$\times PState \times CState$$
$$\to EnergyCost$$

$$repeat'^{ec}(n, cost, body, ps, cs) = \begin{cases} (0 & \text{if } n \leq 0 \\ repeat'^{ec}(n-1, cost, body, ps', cs') & \text{if } n > 0 \\ \quad + cost(ps, cs) & \\ \quad \text{where} & \\ \quad (ps', cs') = body(ps, cs) & \end{cases}$$

Using this $repeat^{ec}$ function, the definition of the rule for the repeat is analogous to the previous definition.

$$\frac{\Delta^{ec} \vdash e : \langle V_{ex}, \Sigma_{ex}, E_{ex} \rangle \qquad \Delta^{ec} \vdash S : \langle \Sigma_{st}, E_{st} \rangle}{\Delta^{ec} \vdash \text{repeat } e \text{ begin } S \text{ end} : \langle \dots, E_{ex} \mp repeat^{ec}(V_{ex}, \Sigma_{ex}, E_{st}, \Sigma_{st}) \rangle} \text{ (etRepeat)}$$

Next is the component function call. The energy cost of a component function call consists of the time taken to execute this function and of explicit energy cost attributed to this call. The environment Δ^{ec} is extended for each component function $C.f$ with two elements: an energy judgment $E_{f'}$ and a run-time $t_{f'}$. Time independent energy usage can be encoded into this $E_{f'}$ function. For functions defined in the language the derived energy judgement is inserted into the environment. There is no need for these functions for an explicit run-time as this is part of the derived energy judgement. Using the patterns described above the component function call is expressed as:

$$\frac{\Delta^{ec} \vdash e : \langle V_{ex}, \Sigma_{ex}, E_{ex} \rangle}{\begin{array}{l} \Delta^{ec}, C.f = (x_{f'}, \dots, \dots, E_{f'}, t_{f'}) \vdash \\ \quad C.f(e) : \langle \dots, \dots, E_{ex} \mp ([x_{f'} \leftarrow V_{ex}, \Sigma_{ex}] \ggg (td^{ec}(t_{f'}) \mp E_{f'})) \rangle \end{array}} \text{ (etCallCmpF)}$$

5   Example

In this section, we demonstrate our analysis on two example programs, comparing their energy usage. Each ECA language construct has an associated execution time bound. These execution times are used in calculating the energy consumption that is time-dependent. Another source of energy consumption in our modelling is time-independent energy usage, which can also be associated with any ECA construct. Adding these two sources together yields the energy consumption.

Consider the example programs in Listings 1 and 2. Both programs play #n beeps at #hz Hz (#n and #hz are input variables). The effect of the first statement is starting a component named SoundSystem, which models a sound card (actually, functions have a single parameter; this parameter is omitted here

```
1  SoundSystem.on();
2  repeat #n begin
3      SoundSystem.playBeepAtHz(#hz);
4      System.sleep()
5  end;
6  SoundSystem.off()
```

Listing 1. Example program

```
1  repeat #n begin
2      SoundSystem.on();
3      SoundSystem.playBeepAtHz(#hz);
4      SoundSystem.off();
5      System.sleep()
6. end
```

Listing 2. Alternative program

as it is not used). After enabling component SoundSystem, beeps are played by calling the function SoundSystem.playBeepAtHz(). Eventually the device is switched off. The sound system has two states, off < on.

For this example, we will assume a start state in which the component is in the off state and an input $n \in \mathbb{Z}$. Furthermore, the time-independent energy cost of SoundSystem.playBeepAtHz() function is i and a call to this function has an execution time of t seconds. The System.sleep() has no associated (time-independent) energy usage, and takes s seconds to complete. The other component function calls and the loop construct are assumed to have zero execution time and zero (time-independent) energy consumption. While switched on, the component has a power draw of u J/s (or W).

We will start with an intuitive explanation of the energy consumption of the program in Listing 1, then continue by applying the analysis presented in this paper and comparing these results. Quickly calculating the execution time of the program yields a result of $n \times (t+s)$ seconds. As the component is switched on at the start and switched off at the end of the program, the time-dependent energy consumption is $n \times (t+s) \times u$. The time-independent energy usage is equal to the number of calls to transmit, thus $n \times i$, resulting in an energy consumption of $n \times (i + (t+s) \times u)$ J.

Now, we will show the results from our analysis. Applying the energy-aware dependent typing rules ultimately yields an energy consumption function and a state update function. Both take a typing and a component state environment as input. The first rule to apply is *etStmtConcat*, with the SoundSystem.on() call as S_1 and the remainder of the program as S_2. The effect of the component function call, calculated with *etCallCmpF*, is that the component state is increased (as this signifies a higher time-dependent energy usage). This effect is represented in the component state update function $\delta_{SoundSystem.on}$. Since we have assumed that the call costs zero time and has no time-independent energy cost, the resulting energy consumption function is the identity function *id*.

We can now analyse the loop with the *etRepeat* rule. We first derive the type of expression e, which determines the number of iterations of the loop. As the expression is a simple variable access, which we assumed not to have any associated costs, the program state and the component states are not touched. For the result of the expression the type system derives $Lookup_{\#n}$ as V_{ex} in the repeat rule, which is (a look-up of) n. This is a function that, when given an input environment, calculates a number in $\mathbb{Z}$ which signifies the number of iterations.

Moving on, we analyse the body of the loop. This means we again apply the $etCallCmpF$ rule to determine the resource consumption of the call to SoundSystem.playBeepAtHz(). The call takes time t. Its energy consumption is u if the component is switched on. For ease of presentation, we here represent the component state $\{off, on\}$ as a variable $e \in \{0, 1\}$. The time-independent energy usage of the loop body is i. The energy consumption function E_{st} of the body is $i + e \times (t + s) \times u$ J.

We can now combine the results of the analysis of the number of iterations and the resource consumption of the loop body (E_{st}) to calculate the consumption of the entire loop. Basically, the resource consumption of the loop E_{st} is multiplied by the number of iterations V_{ex}. This is done in the function $repeat^{ec}(V_{ex}, \Sigma_{ex}, E_{st}, \Sigma_{st})$, in this case $repeat^{ec}(Lookup_{\#n}, id, E_{st}, id)$. Evaluation after the state update function (corresponding to the SoundSystem.on() call), which will update the state of the sound system to on. Evaluating the sequence of both expressions on the start state results in $n \times (i + (t + s) \times u)$ J, signifying the energy consumption of the code fragment.

Analysing the code fragment listed in Listing 2 will result in a type equivalent to $n \times (i + t \times u)$ J, as the cost of switching the sound system on or off is zero. Given the energy characteristics of the sound system, the second code fragment will have a lower energy consumption. Depending on the realistic characteristics (switching a device on or off normally takes time and energy), a realistic trade-off can be made.

6   Related Work

Much of the research that has been devoted to producing green software is focussed on a very abstract level, defining programming and design patterns for writing energy-efficient code [1,8–11]. In [12], a modular design for energy-aware software is presented that is based on a series of rules on UML schemes. In [13,14], a program is divided into "phases" describing similar behaviour. Based on the behaviour of the software, design level optimisations are proposed to achieve lower energy consumption.

Contrary to the abstract levels of the aforementioned papers, there is also research on producing green software on a very low, hardware-specific level [3,15]. Such analysis methods work on specific architectures ([3] on SIMPLESCALAR, [15] on XMOS ISA-level models), while our approach is hardware-parametric.

Furthermore, there is research on building compilers that optimize code for energy-efficiency. In [2], the implementation of several energy optimization methods, including dynamic voltage scaling and bit-switching minimization, into the GCC compiler is described and evaluated. Iterative compilation is applied to energy consumption in [16]. In [17], a technique is proposed in which register labels are encoded alternatively to minimize switching costs. This saves an average of 4.25 % of energy, without affecting performance. The human-written assembly code for an MP3 decoder is optimized by hand in [18], guided by an energy profiling tool based on a proprietary ARM simulator. In [19], functional units are disabled to reduce (leakage) energy consumption of a VLIW processor. The same is done for an adaptation of a DEC Alpha 21264 in [20]. Reduced bit-width instruction set architectures are exploited to save energy in [21]. Energy is saved on memory optimizations in [22–25], while [26,27] focus on variable-voltage processors.

A framework called GREEN is presented in [28]. This framework allows pro-grammers to approximate expensive functions and calculate Quality of Service (QoS) statistics. It can thus help leverage a trade-off between performance and energy consumption on the one hand, and QoS on the other.

In [29], Resource-Utilization Models (RUMs) are presented, which are an abstraction of the resource behaviour of hardware components. The component models in this paper can be viewed as an instance of a RUM. RUMs can be analysed, e.g., with the model checker UPPAAL, while we use a dedicated depen-dent type system. A possible future research direction is to incorporate RUMs into our analysis as component models.

Analyses for consumption of generic resources are built using recurrence rela-tion solving [30], amortized analysis [31], amortization and separation logic [32] and a Vienna Development Method style program logic [33]. The main differences with our work are that we include an explicit hardware model and a context in the form of component states. This context enables the inclusion of power-draw that depends on the state of components..

Several dependently typed programming languages exist, such as EPIGRAM [34] and AGDA [35]. The DEPUTY system, which adds dependent typing to the C language, is described in [36]. Dependent types are applied to security in [37]. Security types there enforce access control and information flow policies.

7 Discussion

The foreseen application area of the proposed analysis is predicting the energy consumption of control systems, in which software controls a number of periph-erals. This includes control systems in factories, cars, airplanes, smart-home applications, etc. Examples of hardware components range from a disk drive or sound card, to heaters, engines, motors and urban lighting. The proposed analy-sis can predict the energy consumption of multiple algorithms and/or different hardware configurations. The choice of algorithm or configuration may depend on the expected workload. This makes the proposed technique useful for both programmers and operators.

The used hardware models can be abstracted, making clear the relative dif-ferences between component-methods and component-states. This makes the proposed approach still applicable even when no final hardware component is available for basing the hardware model on, or this model is not yet created. We observe that many decisions are based on relative properties between systems.

The type system derives precise types, in a syntax-directed manner. Sound-ness and completeness can be proven by induction on the syntax structure of the program. This proof is similar to the proof in [7], but more straightforward due to the absence of approximations.

There are certain properties hardware models must satisfy. Foremost, the models are discrete. Implicit state changes by hardware components cannot be expressed. Energy consumption that gradually increases or decreases over time can therefore not be modelled directly. However, discrete approximations may be used. Compared to the Hoare logic in [5], many restrictions are not present in the proposed type system. Foremost, this type system does not have the limitation that state change cannot depend on the argument of a component function nor that the return value of a component function cannot depend on the state of the component. More realistic models can therefore be used.

The quality of the derived energy expressions is directly related to the quality of the used hardware models. We envision that, in many cases, relative consumption information is sufficient to support design decisions. Depending on the goal, it is possible to use multiple models for one and the same hardware component. For instance, if the hardware model is constructed as a worst-case model, the type system will produce worst-case information. Similarly one can use average-case models to derive average case information.

8   Future Work

In future work, we aim to implement this analysis in order to evaluate its suitability for larger systems and validate practical applicability. We intend to experiment with implementations of various derived approximating analyses in order evaluate which techniques/approximations work best in which context.

A limitation is currently that only a system controlled by *one* central processor can be analysed. Modern systems often consists of a network of interacting systems. Therefore, incorporating interacting systems would increase applicability of the approach.

Another future research direction is to expand the supported input language. Currently, recursion is not supported. An approach to add recursion to the input language is to use the function signatures to compose a set of cost relations (a special case of recurrence relations). A recurrence solver can then eliminate the recursion in the resulting function signatures. In order to support data types, a size analysis of data types is needed to enable iteration over data structures, e.g. using techniques similar to [38, 39].

On the language level the type system is precise, however it does not take into account optimisations and transformations below the language level. This can be achieved by analysing the software on a lower level, for example the intermediate representation of a modern compiler. Another motivation to use such an intermediate representation as the input language is support for (combinations of) many higher level languages. In this way, programs written in and consisting of multiple languages can be analysed. It can also account for optimisations (such as common subexpression elimination, inlining, statically evaluating expressions), which in general reduce the execution time of the program and therefore impact the time-dependent energy usage (calls with side effects like component function calls are generally not optimised).

9 Conclusion

The presented type system captures energy bounds for software that is executed on hardware components of which component models are available. It is *precise, modular and elegant*, yet retains the hardware-parametric aspect.

The new type system is in itself *precise*, as there is no over-estimation, as opposed to the Hoare Logic [5], in which the conditional and loop are over-estimated. Also the class of programs that can be studied is larger, as many of the restrictions needed for the over-approximation are now lifted.

The presented hardware-parametric dependent type system with function signatures enables *modularity*. While analysing the energy consumption of an electronic system, instead of re-analysing the body of functions each time a function call is encountered, the function signature is reused.

By using a dependent type system to express all variables in terms of input variables, the resulting approach is *elegant and concise*, as no externally verified properties are needed.

References

1. Saxe, E.: Power-efficient software. Commun. ACM **53**(2), 44–48 (2010)
2. Zhurikhin, D., Belevantsev, A., Avetisyan, A., Batuzov, K., Lee, S.: Evaluating power aware optimizations within GCC compiler. In: GROW-2009: International Workshop on GCC Research Opportunities (2009)
3. Jayaseelan, R., Mitra, T., Li, X.: Estimating the worst-case energy consumption of embedded software. In: Proceedings of the 12th IEEE Real-Time and Embedded Technology and Applications Symposium, pp. 81–90. IEEE (2006)
4. Ferreira, M.A., Hoekstra, E., Merkus, B., Visser, B., Visser, J.: SEFLab: A lab for measuring software energy footprints. In: 2013 2nd International Workshop on Green and Sustainable Software (GREENS), pp. 30–37. IEEE (2013)
5. Kersten, R., Toldin, P.P., Gastel, B., Eekelen, M.: A hoare logic for energy consumption analysis. In: Dal Lago, U., Peña, R. (eds.) FOPARA 2013. LNCS, vol. 8552, pp. 93–109. Springer, Heidelberg (2014). doi:10.1007/978-3-319-12466-7_6
6. Schoolderman, M., Neutelings, J., Kersten, R., van Eekelen, M.: ECAlogic: hardware-parametric energy-consumption analysis of algorithms. In: Proceedings of the 13th Workshop on Foundations of Aspect-oriented Languages, FOAL 2014, pp. 19–22. ACM, New York (2014)
7. Parisen Toldin, P., Kersten, R., van Gastel, B., van Eekelen, M.: Soundness proof for a hoare logic for energy consumption analysis. Technical Report ICIS-R13009, Radboud University Nijmegen, October 2013
8. Albers, S.: Energy-efficient algorithms. Commun. ACM **53**(5), 86–96 (2010)
9. Ranganathan, P.: Recipe for efficiency: principles of power-aware computing. Commun. ACM **53**(4), 60–67 (2010)
10. Naik, K., Wei, D.S.L.: Software implementation strategies for power-conscious systems. Mob. Netw. Appl. **6**(3), 291–305 (2001)
11. Sivasubramaniam, A., Kandemir, M., Vijaykrishnan, N., Irwin, M.J.: Designing energy-efficient software. In: International Parallel and Distributed Processing Symposium, Los Alamitos, CA, USA. IEEE Computer Society (2002)

12. te Brinke, S., Malakuti, S., Bockisch, C., Bergmans, L., Akşit, M.: A design method for modular energy-aware software. In: Proceedings of the 28th Annual ACM Symposium on Applied Computing, pp. 1180–1182. ACM, New York (2013)

13. Cohen, M., Zhu, H.S., Senem, E.E., Liu, Y.D.: Energy types. SIGPLAN Not. **47**(10), 831–850 (2012)

14. Sampson, A., Dietl, W., Fortuna, E., Gnanapragasam, D., Ceze, L., Grossman, D.: EnerJ: approximate data types for safe and general low-power computation. SIGPLAN Not. **46**(6), 164–174 (2011)

15. Kerrison, S., Liqat, U., Georgiou, K., Mena, A.S., Grech, N., Lopez-Garcia, P., Eder, K., Hermenegildo, M.V.: Energy consumption analysis of programs based on XMOS ISA-level models. In: Gupta, G., Peña, R. (eds.) LOPSTR 2013. LNCS, vol. 8901, pp. 72–90. Springer, Heidelberg (2014)

16. Gheorghita, S.V., Corporaal, H., Basten, T.: Iterative compilation for energy reduction. J. Embedded Comput. **1**(4), 509–520 (2005)

17. Mehta, H., Owens, R.M., Irwin, M.J., Chen, R., Ghosh, D.: Techniques for low energy software. In: ISLPED 1997: Proceedings of the 1997 International Symposium on Low Power Electronics and Design, pp. 72–75. ACM, New York (1997)

18. Šimunić, T., Benini, L., De Micheli, G., Hans, M.: Source code optimization and profiling of energy consumption in embedded systems. In: ISSS 2000: Proceedings of the 13th International Symposium on System Synthesis, Washington, DC, pp. 193–198. IEEE Computer Society (2000)

19. Zhang, W., Kandemir, M., Vijaykrishnan, N., Irwin, M.J., De, V.: Compiler support for reducing leakage energy consumption. In: DATE 2003: Proceedings of the Conference on Design, Automation and Test in Europe, Washington, DC. IEEE Computer Society (2003)

20. You, Y.P., Lee, C., Lee, J.K.: Compilers for leakage power reduction. ACM Trans. Des. Autom. Electron. Syst. **11**(1), 147–164 (2006)

21. Shrivastava, A., Dutt, N.: Energy efficient code generation exploiting reduced bit-width instruction set architectures (rISA). In: ASP-DAC 2004: Proceedings of the 2004 Asia and South Pacific Design Automation Conference, Piscataway, NJ, USA, pp. 475–477. IEEE Press (2004)

22. Joo, Y., Choi, Y., Shim, H., Lee, H.G., Kim, K., Chang, N.: Energy exploration and reduction of SDRAM memory systems. In: DAC 2002: Proceedings of the 39th Annual Design Automation Conference, pp. 892–897. ACM, New York (2002)

23. Verma, M., Wehmeyer, L., Pyka, R., Marwedel, P., Benini, L.: Compilation and simulation tool chain for memory aware energy optimizations. In: SAMOS, pp. 279–288 (2006)

24. Jones, T.M., O'Boyle, M.F.P., Abella, J., González, A., Ergin, O.: Energy-efficient register caching with compiler assistance. ACM Trans. Archit. Code Optim. **6**(4), 1–23 (2009)

25. Lee, H.G., Chang, N.: Energy-aware memory allocation in heterogeneous non-volatile memory systems. In: ISLPED 2003: Proceedings of the 2003 International Symposium on Low Power Electronics and Design, pp. 420–423. ACM, New York (2003)

26. Okuma, T., Yasuura, H., Ishihara, T.: Software energy reduction techniques for variable-voltage processors. IEEE Des. Test **18**(2), 31–41 (2001)

27. Saputra, H., Kandemir, M., Vijaykrishnan, N., Irwin, M.J., Hu, J.S., Hsu, C.H., Kremer, U.: Energy-conscious compilation based on voltage scaling. LCTES/S-COPES 2002: Proceedings of the Joint Conference on Languages, Compilers and Tools for Embedded Systems, pp. 2–11. ACM, New York (2002)

28. Baek, W., Chilimbi, T.M.: Green: a framework for supporting energy-conscious programming using controlled approximation. SIGPLAN Not. **45**(6), 198–209 (2010)
29. te Brinke, S., Malakuti, S., Bockisch, C.M., Bergmans, L.M.J., Akşit, M., Katz, S.: A tool-supported approach for modular design of energy-aware software. In: Proceedings of the 29th Annual ACM Symposium on Applied Computing, Gyeongju, Korea, SAC 2014. ACM, March 2014
30. Albert, E., Arenas, P., Genaim, S., Puebla, G.: Closed-form upper bounds in static cost analysis. J. Autom. Reason. **46**(2), 161–203 (2011)
31. Hoffmann, J., Aehlig, K., Hofmann, M.: Multivariate amortized resource analysis. In: Ball, T., Sagiv, M. (eds.) POPL 2011, pp. 357–370. ACM (2011)
32. Atkey, R.: Amortised resource analysis with separation logic. In: Gordon, A.D. (ed.) ESOP 2010. LNCS, vol. 6012, pp. 85–103. Springer, Heidelberg (2010). doi:10.1007/978-3-642-11957-6_6
33. Aspinall, D., Beringer, L., Hofmann, M., Loidl, H.W., Momigliano, A.: A program logic for resources. Theor. Comput. Sci. **389**(3), 411–445 (2007)
34. McBride, C.: Epigram: practical programming with dependent types. In: Vene, V., Uustalu, T. (eds.) AFP 2004. LNCS, vol. 3622, pp. 130–170. Springer, Heidelberg (2005). doi:10.1007/11546382_3
35. Bove, A., Dybjer, P., Norell, U.: A brief overview of agda - a functional language with dependent types. In: Berghofer, S., Nipkow, T., Urban, C., Wenzel, M. (eds.) TPHOLs 2009. LNCS, vol. 5674, pp. 73–78. Springer, Heidelberg (2009)
36. Condit, J., Harren, M., Anderson, Z., Gay, D., Necula, G.C.: Dependent types for low-level programming. In: Nicola, R. (ed.) ESOP 2007. LNCS, vol. 4421, pp. 520–535. Springer, Heidelberg (2007). doi:10.1007/978-3-540-71316-6_35
37. Morgenstern, J., Licata, D.R.: Security-typed programming within dependently typed programming. Proceedings of the 15th ACM SIGPLAN International Conference on Functional Programming, ICFP 2010, pp. 169–180. ACM, New York (2010)
38. Shkaravska, O., Eekelen, M., Tamalet, A.: Collected size semantics for strict functional programs over general polymorphic lists. In: Dal Lago, U., Peña, R. (eds.) FOPARA 2013. LNCS, vol. 8552, pp. 143–159. Springer, Heidelberg (2014). doi:10.1007/978-3-319-12466-7_9
39. Tamalet, A., Shkaravska, O., van Eekelen, M.: Size analysis of algebraic data types. In: Achten, P., Koopman, P., Morazán, M. (eds.) Trends in Functional Programming, vol. 9 of Trends in Functional Programming, pp. 33–48. Intellect (2009)

Call-by-Value, Elementary Time and Intersection Types

Erika De Benedetti and Simona Ronchi Della Rocca[✉]

Dipartimento di Informatica, Università Degli Studi di Torino,
Corso Svizzera 185, 10149 Torino, Italy
{debenede,ronchi}@di.unito.it

Abstract. We present a type assignment system for the call-by-value
λ-calculus, such that typable terms reduce to normal form in a number
of steps which is elementary in the size of the term itself and in the rank
of the type derivation. Types are built through non-idempotent and non-
associative intersection, and the system is loosely inspired by elementary
affine logic. The result is due to the fact that the lack of associativity
in intersection supplies a notion of stratification of the reduction, which
recalls the similar notion of stratification in light logics.

1 Introduction

This paper fits in the research area on Implicit Computational Complexity, in
particular with the topic of characterizing complexity classes without referring to
a particular machine model. In this setting, we take here an ML-like approach,
whose starting points are on one hand the use of the λ-calculus as abstract
paradigm of (functional) programming languages, and on the other hand the
use of types as a mean to certify program complexity.

Some important results have already been obtained by following this approach; such results are based essentially on the use of light logics [12] having
a bounded cut elimination complexity. A fundamental notion in the light logics
approach is that of *stratification*, by which we informally mean the fact of organ-
ising computation into different strata through a restriction of the "!" modality,
whose duty is to control the duplication of arguments. Starting from the light
logics and exploiting the Curry-Howard isomorphism, some type assignment systems have been designed through the types-as-formulae approach, such that the
languages inherit the quantitative properties of the logic for free. The systems
DLAL (Dual Light Affine Logic) by Baillot and Terui [3] and STA (Soft Type
Assignment) by Gaboardi and Ronchi Della Rocca [11] are examples of such
type assignment systems, based on LAL (Light Affine Logic) [1] and SLL (Soft
Linear Logic) [13] respectively; both characterise polynomial time computations.
A result that is particularly related to the topic of this paper is the design of the
type system ETAS [7], which is based on ELL (Elementary Linear Logic) [8,12];
ETAS assigns types to terms of the λ-calculus in such a way that, if the reduction
is performed in a call-by-value style, the normalisation time is bounded by an

© Springer International Publishing Switzerland 2016
M. van Eekelen and U. Dal Lago (Eds.): FOPARA 2015, LNCS 9964, pp. 40–59, 2016.
DOI: 10.1007/978-3-319-46559-3_3

elementary function depending only on the initial size of the term. Although the light logics supply a very powerful tool for the complexity control, a natural criticism to this approach is that it generates programming languages with very low expressivity; for example, in STA it is not possible to iterate functions, with the unpleasant outcome that writing algorithms becomes quite a difficult exercise.

We are interested in exploring different tools for controlling the complexity, in order to increase the expressive power of the calculus, while preserving the milestones we illustrated before. A natural approach is to take inspiration from intersection types [6], introduced with the aim of increasing the typability power of simple types. The fundamental system is based on two connectives, the arrow ($\rightarrow$) and the intersection ($\wedge$), the latter allowing to assign to a term the type $A \wedge B$, in case the term has both type A and type B. In particular, $\wedge$ can be used as an extended notion of contraction, allowing to merge premises with different types. Intersection types have been extensively used for characterising semantical properties of terms, for example strong normalisation. Moreover, type assignment systems based on intersection types have been designed for reasoning in a finitary way about the interpretation of terms in continuous [16], stable [14] and relational λ-models. Recently it has been noticed that considering intersection without idempotence (i.e., $A \wedge A \neq A$) allows to prove also quantitative properties of terms, as in [5], where the execution time of λ-terms is bounded via non idempotent intersection types, and in [4], where the dimension of the derivation supplies a bound to the number of normalisation steps.

In [10] we explored a possible application of intersection types to complexity, by showing that erasing the associativity property (i.e., $(A \wedge B) \wedge C \neq A \wedge (B \wedge C)$) allows to introduce a sort of stratification mimicking the one of light logics; we then applied this idea in order to implicitly characterise PTIME. To this aim, we defined a type assignment system for the λ-calculus, called STR, characterising PTIME in a sound and complete way by using non-associative intersection types; moreover we proved that STR, which is inspired by SLL, has a strictly stronger typability power than STA, since it gives type to all and only the strong normalising terms. It is interesting to note that in system STR a kind of limited iteration is typable, so increasing the number of algorithms that can be coded with respect to STA. A previous related result is in [9], where a type assignment system using non-idempotent and non-associative intersection is used for proving a bound on the normalisation time of the strongly normalising λ-terms.

In this paper we extend the above mentioned approach to Elementary Time, by designing a type assignment system $\mathcal{M}$, inspired by ELL, where types are built using non-idempotent and non-associative intersection. This is expressed by considering types syntactically as multisets, such that the nesting of the multiset construction supplies a notion of stratification. Like in [7], λ-calculus is considered in its call-by-value version, as introduced by Plotkin [15]: in fact the type assigned to a value (i.e., a variable or an abstraction) reflects in its stratification the number of duplications the value itself can undergo during the reduction, while the naive call-by-name approach causes the subject reduction property to fail, since also applications can have a non-linear type and thus be marked as duplicable, while in general they are not. We prove that this system is sound for elementary time, in the sense that a typed term reduces to a normal form in a number of reductions which is an elementary function of its size and of the rank of its type derivation. Moreover $\mathcal{M}$ is strictly more powerful than ETAS [7], which is based directly on ELL, so enforcing our idea that the approach based on intersection types can be very fruitful for controlling the reduction bound. The use of call-by-value is also motivated by the result in [2], where a complete characterisation of the exponential hierarchy is given, through an extension of the λ-calculus typed by formulae of ELL. Also in this case, λ-calculus has been used in its call-by-value version, since the stratification supplies a natural bound to the types of values.

The paper is organised as follows. Section 2 introduces the type assignment system $\mathcal{M}$ and its basic properties. Section 3 proves a quantitative subject reduction property, which gives a bound to the growth of the subject during the reduction steps. Appendix A contains the very technical proof of the Substitution Lemma. Section 4 contains the main result of the paper, namely the proof that both the number of reduction steps necessary to reduce a typed term to normal form and the size of all its reducts in the normalisation reduction sequence are bound by elementary functions, depending on the size of the term and on the rank of its derivation. Section 5 concludes with some general observations and some hints on future developments.

2   The Type System $\mathcal{M}$

In this section we introduce the call-by-value λ-calculus and its type assignment system $\mathcal{M}$. System $\mathcal{M}$ is in fact inspired by ELL (Elementary Linear Logic), where the modality is replaced by a non-idempotent and non-associative version of the intersection, represented here as multiset union.

Calculus. The language Λ_v is given by the following grammar:

$$M :: = x \mid \lambda x.M \mid MM$$

where $x \in$ Var, a countable set of variables. As usual, terms are considered modulo α-rule, i.e., modulo names of bound variables. The set of values is $V = \{M \in \Lambda \mid M = x$ or $M = \lambda x.N$, for any $x \in$ Var, $N \in \Lambda_v\}$, while the size of a term is defined inductively as $|x| = 1; |\lambda x.M| = |M| + 1; |MN| = |M| + |N| + 1$.

Term contexts are defined as $\mathcal{C} :: = \square \mid \lambda x.\mathcal{C} \mid M\mathcal{C} \mid \mathcal{C}M$, where $\square$ is the hole of the context. $\mathcal{C}[M]$ is the term obtained by plugging term M into the hole of $\mathcal{C}$. Note that, as usual, the plugging operation can cause capture of variables.

The reduction relation $\rightarrow_v$ is the contextual closure of the rule

$$(\lambda x.M)N \rightarrow M[N/x] \quad \text{if and only if } N \in V$$

where $M[N/x]$ denotes the capture-free substitution of all occurrences of x in M by N.

Types. Let $[a_1, \ldots, a_n]$ denote the (unordered) *multiset* of elements $a_1, \ldots, a_n$ and let $[\,]$ denote the empty multiset; moreover, let $\sqcup$ denote the union operator on multisets taking into account multiplicities.

The set $\mathcal{T}$ of types is given by the following grammars:

$$A, B, C \quad ::= a \mid \sigma \to \sigma \qquad \text{(linear types)}$$
$$\sigma, \rho, \mu, \tau ::= A \mid [\sigma_1, \ldots, \sigma_n] \qquad \text{(types)}$$

where a ranges over a countable set of type variables and $n \geq 1$, so the empty multiset is not allowed. A type is *quasi-linear* if it is of the shape $[A_1, \ldots, A_n]$, it is a *multiset type* if it is of the shape $[\sigma_1, \ldots, \sigma_n]$, for some types $\sigma_1, \ldots, \sigma_n$ $(n \geq 1)$; $\mathcal{T}_{\updownarrow}$ and $\mathcal{T}_{\updownarrow}$ denote respectively the set of quasi-linear types and the set of multiset types.

The *type structure* $\mathtt{m}(\sigma)$ is a multiset of pairs $\langle d, A \rangle$, such that $d \in \mathbb{N}$ ($\mathbb{N}$ being the set of natural numbers) and A is a linear type, defined by induction on σ:

- $\mathtt{m}(A) = [\langle 0, A \rangle]$;
- $\mathtt{m}([\sigma_1, \ldots, \sigma_n]) = [\langle k + 1, A \rangle \mid \langle k, A \rangle \in \sqcup_{i=1}^{n} \mathtt{m}(\sigma_i)]$.

The second projection of each pair in $\mathtt{m}(\sigma)$ is a *linear component* of σ.

Equivalences between types and type structures are mutually defined by induction on types and type structures. The equivalence between types is defined as follows:

- $a \cong a$;
- $\rho \cong \rho'$ and $\mu \cong \mu'$ imply $\rho \to \mu \cong \rho' \to \mu'$;
- $\mathtt{m}(\sigma) \cong \mathtt{m}(\tau)$ implies $\sigma \cong \tau$.

while equivalence between type structures is defined as:

- $A \cong A'$ implies $\mathtt{m}(A) \cong \mathtt{m}(A')$;
- $A \cong A'$ and $\mathtt{m}(\rho) \cong \mathtt{m}(\rho')$ imply $[\langle k, A \rangle] \sqcup \mathtt{m}(\rho) \cong [\langle k, A' \rangle] \sqcup \mathtt{m}(\rho')$.

The size of a type σ, denoted by $|\sigma|$, is the cardinality of the multiset $\mathtt{m}(\sigma)$; note that $\|[\sigma_1, \ldots, \sigma_n]\| = \sum_{i=1}^{n} |\sigma_i|$.

Observe that multiset types are allowed on the righthand side of the arrow; this is reminiscent of the (!) rule of ELL, where neither dereliction nor digging are valid principles. Another observation can be made about multiset types, and in particular the equivalence relation defined on them, which takes into account only the number of nested multiset constructions enclosing a linear component: the reason behind this choice is the fact that usual associativity would cause the subject reduction property to fail, whereas the considered equivalence relation allows to maintain the same context and type during a reduction step.

From now on, we consider types modulo $\cong$; it is easy to check that the notion of size is preserved by this relation.

$$\frac{}{\mathtt{x}:[\mathtt{A}]\mid\emptyset\vdash\mathtt{x}:\mathtt{A}}\ (Ax) \qquad \frac{\Gamma\mid\Delta\vdash\mathtt{M}:\sigma\quad(\Gamma\sqcup\Gamma')\#(\Delta\sqcup\Delta')}{\Gamma\sqcup\Gamma'\mid\Delta\sqcup\Delta'\vdash\mathtt{M}:\sigma}\ (w)$$

$$\frac{\Gamma,\mathtt{x}:[\mathtt{A}]\mid\Delta\vdash\mathtt{M}:\sigma}{\Gamma\mid\Delta\vdash\lambda\mathtt{x}.\mathtt{M}:\mathtt{A}\to\sigma}\ (\to I_L) \qquad \frac{\Gamma\mid\Delta,\mathtt{x}:\tau\vdash\mathtt{M}:\sigma}{\Gamma\mid\Delta\vdash\lambda\mathtt{x}.\mathtt{M}:\tau\to\sigma}\ (\to I_M)$$

$$\frac{\Gamma\mid\Delta\vdash\mathtt{M}:\tau\to\sigma\quad\Gamma'\mid\Delta'\vdash\mathtt{N}:\tau\quad(\Gamma\sqcup\Gamma')\#(\Delta\sqcup\Delta')}{\Gamma\sqcup\Gamma'\mid\Delta\sqcup\Delta'\vdash\mathtt{MN}:\sigma}\ (\to E)$$

$$\frac{\Gamma_i\mid\Delta_i\vdash\mathtt{M}:\sigma_i\quad\Gamma_i\#\Delta_j\text{ for every }i,j\quad(1\le i\le n)}{\emptyset\mid\sqcup_{i=1}^n\Gamma_i\sqcup_{i=1}^n[\Delta_i]\vdash\mathtt{M}:[\sigma_1,...,\sigma_n]}\ (m)$$

Fig. 1. The system $\mathcal{M}$.

Contexts. A *context* Γ is a function from variables to $(\mathcal{T}\cup\{[\,]\})/\cong$, such that $\mathrm{dom}(\Gamma)=\{\mathtt{x}\mid\Gamma(\mathtt{x})\ne[\,]\}$ is finite. A context is *linear* if its codomain is restricted to $(\mathcal{T}_{\updownarrow}\cup\{[\,]\})/\cong$, while it is a multiset context if its codomain is restricted to $(\mathcal{T}_{\Updownarrow}\cup\{[\,]\})/\cong$; observe that type contexts are considered modulo $\cong$ by definition.

Contexts are ranged over by Γ,Δ,Θ.

A linear context Γ is *proper* iff, for every $\mathtt{x}\in\mathrm{dom}(\Gamma)$, there is a linear type $\mathtt{A}$ such that $\Gamma(\mathtt{x})=[\mathtt{A}]$.

We denote by $\Gamma\sqcup\Delta$ the union of contexts Γ and Δ, such that $(\Gamma\sqcup\Delta)(\mathtt{x})=\Gamma(\mathtt{x})\sqcup\Delta(\mathtt{x})$. The context $[^k\Gamma]$, such that $\mathrm{dom}([^k\Gamma])=\mathrm{dom}(\Gamma)$, is defined by induction on k as follows: $[^0\Gamma](\mathtt{x})=\Gamma(\mathtt{x})$, while $[^{k+1}\Gamma](\mathtt{x})=[[^k\Gamma](\mathtt{x})]$ and $[\Gamma]$ is short for $[^1\Gamma]$.

A context Γ, such that $\mathrm{dom}(\Gamma)=\{\mathtt{x}_1,...,\mathtt{x}_n\}$ and $\Gamma(\mathtt{x}_i)=\sigma_i$ for every $1\le i\le n$, can be written as a finite sequence $\mathtt{x}_1:\sigma_1,...,\mathtt{x}_n:\sigma_n$. The condition $\Gamma\#\Delta$ denotes the fact that $\mathrm{dom}(\Gamma)\cap\mathrm{dom}(\Delta)=\emptyset$. If $\Gamma\#\Delta$, then $\Gamma\sqcup\Delta$ is written Γ,Δ, and $\Gamma\sqcup_{i=1}^n\Gamma_i$ is short for $\Gamma\sqcup(\sqcup_{i=1}^n\Gamma_i)$.

Example 1. Let $\Gamma_1(\mathtt{x})=[\mathtt{A}]$ and $\Gamma_2(\mathtt{x})=[\mathtt{B}]$: then $(\sqcup_{i=1}^2\Gamma_i)(\mathtt{x})=[\mathtt{A},\mathtt{B}]$ and $[^2\Gamma_1](\mathtt{x})=[^3\mathtt{A}]$.

Type assignment. The type assignment system $\mathcal{M}$, defined in Fig. 1, is a set of rules proving typings of the shape $\Gamma\mid\Delta\vdash\mathtt{M}:\sigma$, where Γ is a linear context, Δ is a multiset context and $\Gamma\#\Delta$. When we need to identify a particular derivation Π proving the typing $\Gamma\mid\Delta\vdash\mathtt{M}:\sigma$, we write $\Pi\triangleright\Gamma\mid\Delta\vdash\mathtt{M}:\sigma$, where $\mathtt{M}$ is the subject of Π.

A few observations about the rules of system $\mathcal{M}$ are in order. The axiom rule allows to introduce only variables having a linear type, while the weakening rule (w) can either introduce a new variable or add some linear components to the type of a variable, which already belongs to the domain of one of the contexts. Rules $(\to I_L)$ and $(\to I_M)$ take into account the context to which the variable to be abstracted belongs: observe that a variable from the linear context can be abstracted only if it is assigned a quasi-linear type of cardinality 1. Rule $(\to E)$ is quite standard, while rule (m) is the key rule of the system and it recalls the rule for the ! modality in Elementary Linear Logic, which is:

$$\frac{\Gamma \vdash \sigma}{!\Gamma \vdash !\sigma} \ (!)$$

In fact, as this rule introduces the modality on both sides of the $\vdash$, our rule builds multisets on both sides; this is achieved by contracting a finite number of derivations with the same subject, such that all types of the same variable are collected into a multiset (inside the multiset context) and the types assigned to the subject become elements of a multiset type; observe that, accordingly, the linear context is transformed into a multiset one.

The condition that the linear and multiset contexts have disjoint domains is implicit in all rules, but rules (w) and $(\to E)$ need the explicit condition that the final contexts will satisfy this requirement. Accordingly, rule (m) imposes that a variable occurring in the linear (resp. multiset) context of a premise does not occur in the multiset (resp. linear) context of another premise. In rule $(\to E)$, if MN is the subject of the rule, we will call the premise with subject M its *function premise*, and the premise with subject N its *argument premise*.

Notice that variables in the linear context whose quasi-linear type is not a singleton cannot be abstracted, hinting to the fact that those variables are in a sort of temporary status, awaiting to be moved to the multiset context; this is reminiscent of the so called parking context of [7]. In particular, we want to avoid abstracting over a variable with multiple occurrences, unless that variables has moved to the multiset context through an application of rule (m). We consider a term M to be typable iff there are Γ, Δ, σ such that $\Gamma \mid \Delta \vdash M : \sigma$ and Γ is proper.

The key property of the system is the following:

Property 1. Let $\Pi \rhd \Gamma \mid \Delta \vdash M : \sigma$, where $M \in V$ and σ is a multiset type; then

i. Π ends by an application of rule (m), followed by a (possibly empty) sequence of applications of rule (w);
ii. there is a derivation proving $\emptyset \mid \Delta \vdash M : \sigma$.

The proof of the property comes directly from the definition of values.

Example 2. We propose a few examples in order to clarify the use of some rules.

1. Rule (w) can either introduce a binding for a variable, which is not yet in the context, or modify the type of a variable, which already belongs to the domain:

$$\frac{\dfrac{}{\mathtt{x} : [\mathtt{A}] \mid \emptyset \vdash \mathtt{x} : \mathtt{A}} \ (Ax)}{\mathtt{x} : [\mathtt{A}], \mathtt{y} : [\mathtt{B}] \mid \emptyset \vdash \mathtt{x} : \mathtt{A}} \ (w) \qquad \frac{\dfrac{}{\mathtt{x} : [\mathtt{A}] \mid \emptyset \vdash \mathtt{x} : \mathtt{A}} \ (Ax)}{\mathtt{x} : [\mathtt{A}, \mathtt{B}] \mid \emptyset \vdash \mathtt{x} : \mathtt{A}} \ (w)$$

2. Rule (m) builds multiset types both on the left and on the right hand side of $\vdash$:

$$\cfrac{\cfrac{\overline{\text{x}:[\text{A}]\mid\emptyset\vdash\text{x}:\text{A}}\ (Ax)\qquad\overline{\text{x}:[\text{B}]\mid\emptyset\vdash\text{x}:\text{B}}\ (Ax)}{\emptyset\mid\text{x}:[\text{A},\text{B}]\vdash\text{x}:[\text{A},\text{B}]}\ (m)}{\Pi\triangleright\emptyset\mid\text{x}:[[\text{A},\text{B}]]\vdash\text{x}:[[\text{A},\text{B}]]}\ (m)$$

3. Observe that the derivation

$$\cfrac{\cfrac{\overline{\text{x}:[\text{A}]\mid\emptyset\vdash\text{x}:\text{A}}\ (Ax)}{\emptyset\mid\text{x}:[\text{A}]\vdash\text{x}:[\text{A}]}\ (m)\qquad\cfrac{\overline{\text{x}:[\text{B}]\mid\emptyset\vdash\text{x}:\text{B}}\ (Ax)}{\emptyset\mid\text{x}:[\text{B}]\vdash\text{x}:[\text{B}]}\ (m)}{\Sigma\triangleright\emptyset\mid\text{x}:[[\text{A}],[\text{B}]]\vdash\text{x}:[[\text{A}],[\text{B}]]}\ (m)$$

is equivalent to the one given in point 2 of the current example, since $\text{m}([[\text{A}],[\text{B}]])=\text{m}([[\text{A},\text{B}]])=[\langle\text{A},2\rangle,\langle\text{B},2\rangle]$ implies $[[\text{A}],[\text{B}]]\cong[[\text{A},\text{B}]]$.

The system $\mathcal{M}$ is strictly more powerful than the similar system ETAS [7], based on Elementary Linear Logic.

Example 3. The self-application $\lambda\text{x.xx}$ is typable in this system, but not in ETAS:

$$\cfrac{\cfrac{\overline{\text{x}:[\text{A}\to\text{A}]\mid\emptyset\vdash\text{x}:\text{A}\to\text{A}}\ (Ax)\qquad\overline{\text{x}:[\text{A}]\mid\emptyset\vdash\text{x}:\text{A}}\ (Ax)}{\cfrac{\text{x}:[\text{A}\to\text{A},\text{A}]\mid\emptyset\vdash\text{xx}:\text{A}}{\emptyset\mid\text{x}:[\text{A}\to\text{A},\text{A}]\vdash\text{xx}:[\text{A}]}\ (m)}\ (\to E)}{\emptyset\mid\emptyset\vdash\lambda\text{x.xx}:[\text{A}\to\text{A},\text{A}]\to[\text{A}]}\ (\to I_M)$$

Note that every linear type for $\lambda\text{x.xx}$ is an arrow type, with multiset types both on the left and on the right hand side of the arrow.

In Sect. 3 we show that all terms typable in $\mathcal{M}$ are strongly normalising; however the system is not complete for strong normalisation: indeed, the term $(\lambda\text{x.xx})(\lambda\text{z.z})(\lambda\text{z.z})$ is not typable in $\mathcal{M}$ (obviously nor in ETAS), despite being a strongly normalising term. The following lemma sketches a proof of this fact.

Lemma 1. *The term* $\text{Q}=(\lambda\text{x.xx})(\lambda\text{z.z})(\lambda\text{z.z})$ *is not typable in* $\mathcal{M}$.

Proof. In order to assign a type to the term Q, we need to assign to its subterm $(\lambda\text{x.xx})(\lambda\text{z.z})$ a linear type of the shape $\sigma\to\tau$ such that σ can be assigned to $(\lambda\text{z.z})$, and so to $(\lambda\text{x.xx})$ a type of the shape $\rho\to\sigma\to\tau$, such that both ρ and σ can be assigned to $(\lambda\text{z.z})$. This would imply having a typing $\Gamma\mid\Delta\vdash\text{xx}:\sigma\to\tau$ where either $\Gamma(\text{x})=[\text{A}]$ and $\rho=\text{A}$, or $\Delta(\text{x})=\rho$, by the condition on rule $(\to I_L)$. The first situation is not possible, since there are two occurrences of x in xx, so two axioms have been used, which implies that $\Gamma(\text{x})$ must be a multiset of cardinality at least two. So it must be $\Delta(\text{x})=\rho$, where ρ is a multiset type. Looking at the previous example, in order to give the type $\sigma\to\tau$ to xx we need to assign to x both an arrow type $\mu\to\sigma\to\tau$ and the type μ. As far as the first assignment is concerned, it is possible only in case x is in the domain of the linear context: then the two premises of x can be contracted only by means of the rule (m), with the consequence that the final type $[\sigma\to\tau]$ is no more useful for an application.

3   Quantitative Properties of $\mathcal{M}$

In this section we prove the subject reduction property for system $\mathcal{M}$, in a quantitative way: namely we enrich the usual subject reduction property by giving a bound on the growth of the subject of the derivation after a reduction step. In order to do this, we need to introduce some quantitative measures of a derivation and to prove that they are preserved by the relation $\cong$.

The **depth** of a derivation Π, denoted by $\mathrm{d}(\Pi)$, is the maximum nesting of applications of rule (m) in it, i.e. the maximum number of applications of rule (m) from any axiom to the root of the derivation. The depth corresponds to the maximum stratification of a derivation, in fact rule (m) is the only rule building multisets types and thus introducing a stratification.

Let Σ be a subderivation of Π, for which we use the notation $\Sigma \in \Pi$: then $\mathrm{d}(\Sigma, \Pi)$ is the number of applications of rule (m) from the root of Σ to the root of Π. It is easy to show that $\mathrm{d}(\Pi) = \max_{\Sigma \in \Pi} \mathrm{d}(\Sigma, \Pi)$.

The **rank** of a derivation Π at a given depth j, denoted by $\mathrm{R}(\Pi, j)$, measures the number of axioms in it at that depth, and it supplies a bound to the size of a type that can be assigned to a value; it is defined by induction on Π as follows:

- if Π is an application of rule (Ax), then $\mathrm{R}(\Pi, 0) = 1$ and $\mathrm{R}(\Pi, j) = 0$ for every $j \geq 1$;
- If Π ends with an application of rule (w), $(\to I_L)$ or $(\to I_M)$ to the premise Σ, then $\mathrm{R}(\Pi, j) = \mathrm{R}(\Sigma, j)$ for every $j \geq 0$;
- if Π ends with an application of rule $(\to E)$ to the premises Σ_1 and Σ_2, then $\mathrm{R}(\Pi, j) = \mathrm{R}(\Sigma_1, j) + \mathrm{R}(\Sigma_2, j)$ for every $j \geq 0$;
- if Π ends with an application of rule (m) to the premises Π_i $(1 \leq i \leq n)$, then $\mathrm{R}(\Pi, 0) = 0$ and $\mathrm{R}(\Pi, j + 1) = \sum_{i=1}^{n} \mathrm{R}(\Pi_i, j)$ for every $j \geq 0$.

The overall rank of Π is $\mathrm{R}(\Pi) = \sum_{j=0}^{\mathrm{d}(\Pi)} \mathrm{R}(\Pi, j)$.

Property 2. Let $\Pi \rhd \Gamma \mid \Delta \vdash \mathsf{M} : \sigma$; then:

  i. $\mathrm{R}(\Pi, j) = 0$ for every $j > \mathrm{d}(\Pi)$;
 ii. $\mathrm{R}(\Pi) \geq 1$;
iii. if $\mathsf{M} \in \mathsf{V}$, then $|\sigma| \leq \mathrm{R}(\Pi)$.

The proof of each point follows easily by induction on Π.

Finally, the **weight** of a derivation Π at a given depth j, denoted by $\mathrm{W}(\Pi, j)$, measures the portion of the subject which lies at that depth; it is defined by induction on Π as follows:

- if Π is an application of rule (Ax), then $\mathrm{W}(\Pi, 0) = 1$ and $\mathrm{W}(\Pi, j) = 0$ for every $j > 0$;
- if Π ends with an application of rule (w) to the premise Σ, then $\mathrm{W}(\Pi, j) = \mathrm{W}(\Sigma, j)$ for every $j \geq 0$;
- if Π ends with an application of rule $(\to I_L)$ or $(\to I_M)$ to the premises Σ, then $\mathrm{W}(\Pi, 0) = \mathrm{W}(\Sigma, 0) + 1$ and $\mathrm{W}(\Pi, j) = \mathrm{W}(\Sigma, j)$;

– if Π ends with an application of rule $(\to E)$ to the premises Σ_1 and Σ_2, then $\mathtt{W}(\Pi, 0) = \mathtt{W}(\Sigma_1, 0) + \mathtt{W}(\Sigma_2, 0) + 1$ and $\mathtt{W}(\Pi, j) = \mathtt{W}(\Sigma_1, j) + \mathtt{W}(\Sigma_2, j)$, for every $j > 0$;

– if Π ends with an application of rule (m) to the premises $\Pi_1, \dots, \Pi_n$, then $\mathtt{W}(\Pi, 0) = 0$ and $\mathtt{W}(\Pi, j+1) = \sum_{i=1}^{n} \frac{|\sigma_i|}{|\sigma|} \cdot \mathtt{W}(\Pi_i, j)$ for every $j \geq 0$, where σ_i, σ are respectively the types of Π_i, Π $(1 \leq i \leq n)$.

The overall weight of Π is $\mathtt{W}(\Pi) = \sum_{j=0}^{\mathtt{d}(\Pi)} \mathtt{W}(\Pi, j)$.

Example 4. Consider the following derivation Π

$$
\cfrac{
\cfrac{
\cfrac{\overline{\mathtt{x} : [A] \mid \emptyset \vdash \mathtt{x} : A}\ (Ax) \qquad \overline{\mathtt{x} : [B] \mid \emptyset \vdash \mathtt{x} : B}\ (Ax)}
{\emptyset \mid \mathtt{x} : [A, B] \vdash \mathtt{x} : [A, B]}\ (m)
\qquad
\overline{\mathtt{x} : [C] \mid \emptyset \vdash \mathtt{x} : C}\ (Ax)
}
{\emptyset \mid \mathtt{x} : [[A, B], C] \vdash \mathtt{x} : [[A, B], C]}\ (m)
}
{}
$$

where $\mathtt{d}(\Pi) = 2$. Here $\mathtt{W}(\Pi, 0) = 0$, $\mathtt{W}(\Pi, 1) = \frac{1}{3}$ and $\mathtt{W}(\Pi, 2) = \frac{2}{3}$, so that $\mathtt{W}(\Pi) = \sum_{j=0}^{2} \mathtt{W}(\Pi, j) = 1$.

Note that the weight of a derivation at depth j may be a rational number, but the sum of the weights over all depths is in fact a natural number. Indeed, it is easy to check that the sum of the weights up to the maximum depth of a derivation Π corresponds to the size of the subject of Π:

Property 3. $\Pi \triangleright \Gamma \mid \Delta \vdash \mathtt{M} : \sigma$ implies $\mathtt{W}(\Pi) = |\mathtt{M}|$.

The rules which contribute in building the subject, namely all rules but (w) and (m), are called *constructive rules*. Observe that, whenever a derivation Π ends with a non-constructive rule, we can always identify a number of **constructive subderivations** of Π whose last rule is constructive; those are defined by induction on Π as follows:

– if the last rule of Π is a constructive rule, then the only constructive subderivation of Π is Π itself;

– if the last rule of Π is either (w) or (m) with premises $\Pi_1, \dots, \Pi_n$, then the constructive subderivations of Π are all the constructive subderivations of $\Pi_1, \dots, \Pi_n$.

We can now extend the equivalence relation $\cong$ to derivations proving the same statement, so that $\Pi \cong \Sigma$ is defined by induction on Π as follows:

– if Π ends with an application of rule (Ax), then $\Pi \cong \Sigma$ if Π and Σ are the same derivation;

– if Π ends with an application of a rule $R \in \{(\to I_L), (\to I_M)\}$ to the premise Π', then $\Pi \cong \Sigma$ if Σ ends with an application of rule R to the premise Σ' and $\Pi' \cong \Sigma'$;

– if Π ends with an application of rule $(\to E)$ to function premise Π_1 and argument premise Π_2, then $\Pi \cong \Sigma$ if Σ ends with an application of rule $(\to E)$ to function premise Σ_1 and argument premise Σ_2, such that $\Pi_1 \cong \Sigma_1$ and $\Pi_2 \cong \Sigma_2$;

– if $\Pi_1 \ldots \Pi_n$ are the constructive subderivations of Π, then $\Pi \cong \Sigma$ if $\Sigma_1 \ldots \Sigma_n$ are the constructive subderivations of Σ and $\Pi_i \cong \Sigma_{j_i}$ for some permutation $\{j_1, \ldots, j_n\}$ of $\{1, \ldots, n\}$.

The following lemma, quite technical, connects the measures of a derivation with the measures of its constructive subderivations (points i and ii) and proves (in point iii) that the equivalence between derivations preserves the measures.

Lemma 2

i. Let $\Pi_i \triangleright \Gamma_i \mid \Delta_i \vdash \mathtt{M} : \mathtt{A}_i$ for $1 \leq i \leq n$; then, for every σ such that $\mathtt{m}(\sigma) = [\langle d_i, \mathtt{A}_i \rangle \mid 1 \leq i \leq n]$ and $I_j = \{i \mid j \geq d_i\}$, there is $\Pi \triangleright \Gamma \mid \Delta \vdash \mathtt{M} : \sigma$ such that $\Pi_1, \ldots, \Pi_n$ are the constructive subderivations of Π, $\mathtt{W}(\Pi, j) = \sum_{i \in I_j} \frac{1}{|\sigma|} \cdot \mathtt{W}(\Pi_i, j - d_i)$ and $\mathtt{R}(\Pi, j) = \sum_{i \in I_j} \mathtt{R}(\Pi_i, j - d_i)$ for every $j \geq 0$.

ii. Let $\Pi \triangleright \Gamma \mid \Delta \vdash \mathtt{M} : \sigma$, where $\mathtt{M} \in V$, $\mathtt{m}(\sigma) = [\langle d_i, \mathtt{A}_i \rangle \mid 1 \leq i \leq n]$ and $I_j = \{i \mid j \geq d_i\}$; then there are $\Pi_i \triangleright \Gamma_i \mid \Delta_i \vdash \mathtt{M} : \mathtt{A}_i$, for $1 \leq i \leq n$, such that $\Pi_1, \ldots, \Pi_n$ are the constructive subderivations of Π, $\mathtt{W}(\Pi, j) = \sum_{i \in I_j} \frac{1}{|\sigma|} \cdot \mathtt{W}(\Pi_i, j - d_i)$ and $\mathtt{R}(\Pi, j) = \sum_{i \in I_j} \mathtt{R}(\Pi_i, j - d_i)$ for every $j \geq 0$.

iii. Let $\Pi \cong \Sigma$. Then $\mathtt{d}(\Pi) = \mathtt{d}(\Sigma)$, $\mathtt{R}(\Pi, j) = \mathtt{R}(\Sigma, j)$ and $\mathtt{W}(\Pi, j) = \mathtt{W}(\Sigma, j)$, for all $j \geq 0$.

Proof. Point i can be proved by induction on σ, point ii by induction on Π, point iii follows from the previous ones using the definition of $\cong$.

Example 5. Consider derivations Π and Σ of Example 2.2 and 2.3 respectively; it is easy to check that:

– $\mathtt{W}(\Pi, j) = \mathtt{W}(\Sigma, j) = 0$ for $j \neq 2$ and $\mathtt{W}(\Pi, 2) = \mathtt{W}(\Sigma, 2) = 1$;
– $\mathtt{R}(\Pi, j) = \mathtt{R}(\Sigma, j) = 0$ for $j \neq 2$ and $\mathtt{R}(\Pi, 2) = \mathtt{R}(\Sigma, 2) = 2$;

Accordingly, $\mathtt{W}(\Pi) = 1 = |\mathtt{x}|$ (by Property 3) and $\mathtt{R}(\Pi) = 2 = |[[\mathtt{A}, \mathtt{B}]]|$ (by Property 2.iii).

We can now state the Substitution Lemma, distinguishing between the cases of the substituted variable occurring in the linear context or in the multiset context, and give a bound to the increase in both weight and rank at every depth.

Lemma 3 (Substitution)

i. Let $\Pi \triangleright \Gamma, \mathtt{x} : [\mathtt{A}_1, \ldots, \mathtt{A}_n] \mid \Delta \vdash \mathtt{M} : \sigma$ and $\Sigma_i \triangleright \Gamma_i \mid \Delta_i \vdash \mathtt{N} : \mathtt{A}_i$, where $\Gamma \# \Delta_i$ and $\Delta \# \Gamma_i$ for $1 \leq i \leq n$: then $\mathcal{S}_\Pi^{\Sigma_1, \ldots, \Sigma_n} \triangleright \Gamma \sqcup_{i=1}^n \Gamma_i \mid \Delta \sqcup_{i=1}^n \Delta_i \vdash \mathtt{M}[\mathtt{N}/\mathtt{x}] : \sigma$ such that $\mathtt{W}(\mathcal{S}_\Pi^{\Sigma_1, \ldots, \Sigma_n}, j) \leq \mathtt{W}(\Pi, j) + \sum_{i=1}^n \mathtt{W}(\Sigma_i, j)$ and $\mathtt{R}(\mathcal{S}_\Pi^{\Sigma_1, \ldots, \Sigma_n}, j) \leq \mathtt{R}(\Pi, j) + \sum_{i=1}^n \mathtt{R}(\Sigma_i, j)$ for every $j \geq 0$.

ii. Let $\Pi \triangleright \Gamma \mid \Delta, \mathtt{x} : \tau \vdash \mathtt{M} : \sigma$ and $\Sigma \triangleright \Gamma' \mid \Delta' \vdash \mathtt{N} : \tau$, where $\mathtt{N} \in V$, $\Gamma \# \Delta'$ and $\Delta \# \Gamma'$: then $\mathcal{S}_\Pi^\Sigma \triangleright \Gamma \sqcup \Gamma' \mid \Delta \sqcup \Delta' \vdash \mathtt{M}[\mathtt{N}/\mathtt{x}] : \sigma$ such that $\mathtt{W}(\mathcal{S}_\Pi^\Sigma, j) \leq \mathtt{W}(\Pi, j) + |\tau| \cdot \mathtt{W}(\Sigma, j)$ and $\mathtt{R}(\mathcal{S}_\Pi^\Sigma, j) \leq \mathtt{R}(\Pi, j) + \mathtt{R}(\Sigma, j)$ for every $j \geq 0$.

Proof. In order to improve readability, the proof, which is quite long, can be found in the Appendix.

Let Π be a derivation typing $\mathcal{C}[\mathtt{M}]$; actually, while the context denotes a unique occurrence of a subterm $\mathtt{M}$ in $\mathcal{C}[\mathtt{M}]$, in Π there might be more than one subderivation typing $\mathtt{M}$, namely many virtual copies of the same redex, called **cuts**, all of which built by subderivations of one of the following shapes:

$$\cfrac{\cfrac{\dfrac{\Psi \rhd \Gamma' \mid \Delta' \vdash \mathtt{P} : \sigma}{\Gamma'' \mid \Delta'' \vdash \lambda\mathtt{x}.\mathtt{P} : \tau \to \sigma}\ (R)}{\Gamma \mid \Delta \vdash \lambda\mathtt{x}.\mathtt{P} : \tau \to \sigma}\ (w) \qquad \Phi \rhd \Theta \mid \Xi \vdash \mathtt{N} : \tau}{\Gamma \sqcup \Theta \mid \Delta \sqcup \Xi \rhd (\lambda\mathtt{x}.\mathtt{P})\mathtt{N} : \sigma}\ (\to E)$$

where either R is $(\to I_L)$, $\Gamma' = \Gamma''$, $\mathtt{x} : [\mathtt{A}]$, $\tau = \mathtt{A}$ and $\Delta' = \Delta''$, or R is $(\to I_M)$, $\Gamma' = \Gamma''$, $\Delta' = \Delta''$, $\mathtt{x} : \tau$.

By Lemma 3, a cut can be eliminated by replacing it by the subderivation $\mathcal{S}_\Psi^\Phi \rhd \Gamma' \sqcup \Theta \mid \Delta' \sqcup \Xi \vdash \mathtt{P}[\mathtt{N}/\mathtt{x}] : \sigma$, followed by some applications of rule (w). Then we can define a notion of reduction of derivations: a reduction step on the subject of a derivation Π corresponds to eliminate simultaneously in Π all the cuts whose subject is the redex to be reduced. We associate to every reduction step a depth, which corresponds to the minimal depth between all cuts whose subject is a virtual copy of the redex to be reduced; we denote by $\mathtt{M} \to_i \mathtt{M}'$ the fact that $\mathtt{M} \to_v \mathtt{M}'$ by reducing a redex at depth i.

Example 6. Consider the following derivation Π:

$$\cfrac{\dfrac{\Phi \rhd \Theta \mid \Xi \vdash (\lambda\mathtt{x}.\mathtt{P})\mathtt{N} : \mathtt{A}}{\emptyset \mid \Theta, [\Xi] \vdash (\lambda\mathtt{x}.\mathtt{P})\mathtt{N} : [\mathtt{A}]}\ (m) \qquad \Sigma \rhd \Gamma \mid \Delta \vdash (\lambda\mathtt{x}.\mathtt{P})\mathtt{N} : \mathtt{B}}{\emptyset \mid \Gamma \sqcup [\Delta] \sqcup [\Theta] \sqcup [[\Xi]] \vdash (\lambda\mathtt{x}.\mathtt{P})\mathtt{N} : [[\mathtt{A}], \mathtt{B}]}\ (m)$$

such that both Φ and Σ are cuts, and let Φ' and Σ' be the derivation obtained by eliminating such cuts according to the procedure described above. Then the derivation for $\mathtt{P}[\mathtt{N}/\mathtt{x}]$ is

$$\cfrac{\dfrac{\Phi' \rhd \Theta \mid \Xi \vdash \mathtt{P}[\mathtt{N}/\mathtt{x}] : \mathtt{A}}{\emptyset \mid \Theta, [\Xi] \vdash \mathtt{P}[\mathtt{N}/\mathtt{x}] : [\mathtt{A}]}\ (m) \qquad \Sigma' \rhd \Gamma \mid \Delta \vdash \mathtt{P}[\mathtt{N}/\mathtt{x}] : \mathtt{B}}{\emptyset \mid \Gamma \sqcup [\Delta] \sqcup [\Theta] \sqcup [[\Xi]] \vdash \mathtt{P}[\mathtt{N}/\mathtt{x}] : [[\mathtt{A}], \mathtt{B}]}\ (m)$$

Note that $\mathtt{d}(\Phi, \Pi) = 2$ and $\mathtt{d}(\Sigma, \Pi) = 1$, so Σ is the cut of lowest depth: then $(\lambda\mathtt{x}.\mathtt{P})\mathtt{N} \to_1 \mathtt{P}[\mathtt{N}/\mathtt{x}]$.

The weighted subject reduction gives in its proof a formal definition of the reduction on derivations, and it takes into consideration the changes in both weight and rank when firing a reduction at a given depth i.

Theorem 1 (Weighted subject reduction). *Let* $\Pi \rhd \Gamma \mid \Delta \vdash \mathtt{M} : \sigma$ *and* $\mathtt{M} \to_i \mathtt{M}'$; *then* $\Pi' \rhd \Gamma \mid \Delta \vdash \mathtt{M}' : \sigma$, *where* $\mathtt{R}(\Pi', j) < \mathtt{R}(\Pi, j)$ *for every* $j \geq 0$ *and*

$$\begin{aligned} \mathtt{W}(\Pi', j) &= \mathtt{W}(\Pi, j) && \text{for } j < i \\ \mathtt{W}(\Pi', i) &< \mathtt{W}(\Pi, i) && \\ \mathtt{W}(\Pi', j) &\leq \mathtt{W}(\Pi, j) \cdot \textstyle\sum_{h=i}^{\delta(\Pi)} \mathtt{R}(\Pi, h) && \text{for } j > i \end{aligned}$$

Proof. Let $M = \mathcal{C}[(\lambda x.P)N]$ and $M' = \mathcal{C}[P[N/x]]$, where $N \in V$; the proof is by induction on $\mathcal{C}$.

Let $\mathcal{C} = \square$: we proceed by induction on Π. Note that the last rule of Π is either (w), $(\to E)$ or (m).

- If the last rule is (w), then the proof follows by induction.
- Let Π end with an application of rule $(\to E)$, so $i = 0$; there are two cases.
  - In the first case, let Π be

$$\cfrac{\cfrac{\cfrac{\Psi \triangleright \Gamma'', x : [A] \mid \Delta'' \vdash P : \sigma}{\Gamma'' \mid \Delta'' \vdash \lambda x.P : A \to \sigma} (\to I_L)}{\Gamma \mid \Delta \vdash \lambda x.P : A \to \sigma} (w) \qquad \Phi \triangleright \Theta \mid \Xi \vdash N : A}{\Gamma \sqcup \Theta \mid \Delta \sqcup \Xi \triangleright (\lambda x.P)N : \sigma} (\to E)$$

Since N is a value, by Lemma 3.i, $\mathcal{S}_\Psi^\Phi \triangleright \Gamma'' \sqcup \Theta \mid \Delta'' \sqcup \Xi \vdash P[N/x] : \sigma$, and $\Pi' \triangleright \Gamma \sqcup \Theta \mid \Delta \sqcup \Xi \vdash P[N/x] : \sigma$ is obtained from $\mathcal{S}_\Psi^\Phi$ by applying rule (w). At depth i we have $W(\Pi', 0) = W(\mathcal{S}_\Psi^\Phi, 0) \leq W(\Psi, 0) + W(\Phi, 0) < W(\Pi, 0)$, while at depth $j > i$ we have $W(\Pi', j) = W(\mathcal{S}_\Psi^\Phi, j) \leq W(\Psi, j) + W(\Phi, j) = W(\Pi, j) \leq \sum_{h=0}^{d(\Pi)} R(\Pi, h) \cdot W(\Pi, j)$. Moreover $R(\Pi', j) = R(\mathcal{S}_\Psi^\Phi, j) \leq R(\Psi, j) + R(\Phi, j) = R(\Pi, j)$.
  - In the second case, let Π be

$$\cfrac{\cfrac{\cfrac{\Psi \triangleright \Gamma'' \mid \Delta'', x : \tau \vdash P : \sigma}{\Gamma'' \mid \Delta'' \vdash \lambda x.P : \tau \to \sigma} (\to I_M)}{\Gamma \mid \Delta \vdash \lambda x.P : \tau \to \sigma} (w) \qquad \Phi \triangleright \Theta \mid \Xi \vdash N : \tau}{\Gamma \sqcup \Theta \mid \Delta \sqcup \Xi \triangleright (\lambda x.P)N : \sigma} (\to E)$$

Since N is a value, by Lemma 3.ii, there is $\mathcal{S}_\Psi^\Phi \triangleright \Gamma'' \sqcup \Theta \mid \Delta'' \sqcup \Xi \vdash P[N/x] : \sigma$, and $\Pi' \triangleright \Gamma \sqcup \Theta \mid \Delta \sqcup \Xi \vdash P[N/x] : \sigma$ is obtained from $\mathcal{S}_\Psi^\Phi$ by applying rule (w). At depth i we have $W(\Pi', 0) = W(\mathcal{S}_\Psi^\Phi, 0) \leq W(\Psi, 0) + |\tau| \cdot W(\Phi, 0) = W(\Psi, 0) < W(\Pi, 0)$, since $W(\Phi, 0) = 0$ by Lemma 5.ii, while at depth $j > i$, $W(\Pi', j) \leq W(\Psi, j) + |\tau| \cdot W(\Phi, j) \leq |\tau| \cdot (W(\Psi, j) + W(\Phi, j)) \leq |\tau| \cdot W(\Pi, j)$ So, by by Property 2.iii, $W(\Pi', j) \leq \sum_{h=0}^{d(\Phi)} R(\Phi, h) \cdot W(\Pi, j) \leq \sum_{h=0}^{d(\Pi)} R(\Pi, h) \cdot W(\Pi, j)$ Moreover $R(\Pi', j) \leq R(\Psi, j) + R(\Phi, j) = R(\Pi, j)$.
- Let Π be

$$\cfrac{\Pi_s \triangleright \Gamma_s \mid \Delta_s \vdash M : \sigma_s \quad (1 \leq s \leq p)}{\emptyset \mid \sqcup_{s=1}^p \Gamma_s, \sqcup_{s=1}^p [\Delta_s] \vdash M : [\sigma_1, \ldots, \sigma_s]} (m)$$

so $\sigma = [\sigma_1, \ldots, \sigma_p]$.
If $(\lambda x.P)N \to_{i_s} N[P/x]$, by inductive hypothesis there is $\Pi'_s \triangleright \Gamma_s \mid \Delta_s \vdash P[N/x] : \sigma_s$, where, for $j < i_s$, $W(\Pi'_s, j) = W(\Pi_s, j)W(\Pi'_s, i_s) < W(\Pi_s, i_s)$. For $j > i_s$, we have $W(\Pi'_s, j) \leq W(\Pi_s, j) \cdot \sum_{h=i_s}^{\delta(\Pi_s)} R(\Pi_s, h)$ and $R(\Pi'_s, j) \leq R(\Pi_s, j)$ for every $j \geq 0$, for $1 \leq s \leq p$.
Note that Π' is obtained by applying rule (m) to $\Pi'_1, \ldots, \Pi'_p$ and $i = \min_{s=1}^p i_s + 1$. For $j < i$, we have $W(\Pi', j) = \sum_{s=1}^p \frac{|\sigma_s|}{|\sigma|} \cdot W(\Pi'_s, j-1) = $

$\sum_{s=1}^{p} \frac{|\sigma_s|}{|\sigma|} \cdot \mathtt{W}(\Pi_s, j-1) = \mathtt{W}(\Pi, j)$ since $j - 1 < i_s$ for every $1 \leq s \leq p$.

For $j = i$, we have $\mathtt{W}(\Pi', j) = \sum_{s=1}^{p} \frac{|\sigma_s|}{|\sigma|} \cdot \mathtt{W}(\Pi'_s, j-1) < \sum_{s=1}^{p} \frac{|\sigma_s|}{|\sigma|} \cdot \mathtt{W}(\Pi_s, j-1) = \mathtt{W}(\Pi, j)$ since $\mathtt{W}(\Pi'_s, j-1) < \mathtt{W}(\Pi_s, j-1)$ if $i_s = \min_{s=1}^{p} i_s$ and $\mathtt{W}(\Pi'_s, i-1) = \mathtt{W}(\Pi_s, i-1)$ if $i_s > \min_{s=1}^{p} i_s$, for $1 \leq s \leq p$.

If $j > i$, we have $\mathtt{W}(\Pi', j) = \sum_{s=1}^{p} \frac{|\sigma_s|}{|\sigma|} \cdot \mathtt{W}(\Pi'_s, j-1) \leq \sum_{s=1}^{p} \frac{|\sigma_s|}{|\sigma|} \cdot \mathtt{W}(\Pi_s, j-1) \cdot \sum_{h_s = i_s}^{\mathtt{d}(\Pi_s)} \mathtt{R}(\Pi_s, h_s) \leq \sum_{s=1}^{p} \frac{|\sigma_s|}{|\sigma|} \cdot \mathtt{W}(\Pi_s, j-1) \cdot \sum_{h=i-1}^{\mathtt{d}(\Pi)-1} \sum_{s=1}^{p} \mathtt{R}(\Pi_s, h) \leq \mathtt{W}(\Pi, j) \cdot \sum_{h=i}^{\mathtt{d}(\Pi)} \mathtt{R}(\Pi, h)$ by Property 2.i. Moreover $\mathtt{R}(\Pi', 0) = \mathtt{R}(\Pi, 0) = 0$ and $\mathtt{R}(\Pi', j+1) = \sum_{s=1}^{p} \mathtt{R}(\Pi'_s, j) \leq \sum_{s=1}^{p} \mathtt{R}(\Pi_s, j) = \mathtt{R}(\Pi, j+1)$ for every $j \geq 0$.

The cases of $\mathcal{C} = \lambda y. \mathcal{C}'$, $\mathcal{C} = \mathtt{Q}\mathcal{C}'$ and $\mathcal{C} = \mathcal{C}'\mathtt{Q}$ follow easily by induction.

Let $\Pi \triangleright \Gamma \mid \Delta \vdash \mathtt{M} : \sigma$, and $\mathtt{M} \rightarrow_i \mathtt{M}'$, and let $\Pi' \triangleright \Gamma \mid \Delta \vdash \mathtt{M}' : \sigma$ be the derivation described by Theorem 1; then we write $\Pi \rightarrow_i \Pi'$.

Example 7. Consider Π and Π' of Example 6: then $\Pi \rightarrow_1 \Pi'$.

Let us point out that the subject reduction does not hold for the usual β-reduction, corresponding to a call-by-name evaluation. In fact it is easy to check that there are two derivations proving respectively: $\mathtt{u} : [\mathtt{E} \rightarrow [\mathtt{A}, \mathtt{B}]], \mathtt{w} : [\mathtt{E}] \mid \emptyset \vdash \mathtt{uw} : [\mathtt{A}, \mathtt{B}]$ and $\emptyset \mid \mathtt{y} : [\mathtt{A} \rightarrow \mathtt{B} \rightarrow \mathtt{C}] \vdash \lambda \mathtt{x}.\mathtt{yxx} : [\mathtt{A}, \mathtt{B}] \rightarrow [\mathtt{C}]$ so we can build a derivation proving: $\mathtt{u} : [\mathtt{E} \rightarrow [\mathtt{A}, \mathtt{B}]], \mathtt{w} : [\mathtt{E}] \mid \mathtt{y} : [\mathtt{A} \rightarrow \mathtt{B} \rightarrow \mathtt{C}] \vdash (\lambda \mathtt{x}.\mathtt{yxx})(\mathtt{uw}) : [\mathtt{C}]$. But it is easy to check that a derivation proving $\mathtt{u} : [\mathtt{E} \rightarrow [\mathtt{A}, \mathtt{B}]], \mathtt{w} : [\mathtt{E}] \mid \mathtt{y} : [\mathtt{A} \rightarrow \mathtt{B} \rightarrow \mathtt{C}] \vdash \mathtt{y}(\mathtt{uw})(\mathtt{uw}) : [\mathtt{C}]$ does not exist, since every derivation with subject $\mathtt{y}(\mathtt{uw})(\mathtt{uw})$ needs to have the premises for $\mathtt{u}$ and $\mathtt{w}$ in the multiset context. This counter-example is the same as the one given in [7], showing that ETAS is not sound for β-reduction.

4   Complexity

Now we are ready to prove the main result of the paper, namely the fact that $\Pi \triangleright \Gamma \mid \Delta \vdash \mathtt{M} : \sigma$ for some σ implies that both the number of the reduction steps from $\mathtt{M}$ to normal form and the size of the reducts of $\mathtt{M}$ are bounded by elementary functions depending only on $|\mathtt{M}|$ and on the rank of Π, which is related to its size. The result is based on Theorem 1 and on the fact that the depth of a derivation does not increase after a reduction; indeed this is a crucial property also of Elementary Linear Logic, by which our system is inspired.

By Theorem 1 a reduction step at depth i leaves unchanged the measures at depth less than i, therefore a depth-by-depth reduction strategy is the less efficient one. Such strategy can be informally defined as follows: given a derivation $\Pi \triangleright \Gamma \mid \Delta \vdash \mathtt{M} : \sigma$, a depth-by-depth reduction strategy is any strategy reducing a redex at depth i in $\mathtt{M}$ only if there are not redexes at depth lower than i. Consider a sequence of derivations obtained by a depth-by-depth reduction strategy:

$$\Pi \underbrace{\rightarrow_0 \ldots \rightarrow_0}_{n_0} \Pi_0 \underbrace{\rightarrow_1 \ldots \rightarrow_1}_{n_1} \Pi_1 \ldots \Pi_i \underbrace{\rightarrow_i \ldots \rightarrow_i}_{n_i} \ldots$$

where Π_i has no cuts at depth less than or equal to i and n_i is the number of reduction steps at depth i, for every $i \leq \mathsf{d}(\Pi)$:

Lemma 4. *There are elementary functions* $f_d, g_d : \mathbb{N} \times \mathbb{N} \to \mathbb{N}$ *such that, for every* $\Pi \triangleright \Gamma \mid \Delta \vdash \mathsf{M} : \sigma$

i. $\sum_{e=0}^{d} \mathsf{W}(\Pi_i, e) \leq f_d(\mathsf{W}(\Pi), \mathsf{R}(\Pi))$;

ii. $\sum_{e=0}^{d} n_e \leq g_d(\mathsf{W}(\Pi), \mathsf{R}(\Pi))$.

Proof. By definition, the subject of Π_i does not contain any redex at depth less than or equal to i. Consider Π_0, i.e. the derivation obtained from Π by reducing all redexes at depth 0; by Theorem 1, $\mathsf{W}(\Pi_0, 0) < \mathsf{W}(\Pi, 0)$ and $\mathsf{W}(\Pi_0, j) \leq \mathsf{W}(\Pi, j) \cdot (\mathsf{R}(\Pi))^{n_0}$ for $j > 0$; moreover $n_0 \leq \mathsf{W}(\Pi, 0)$ since the number of reduction at depth 0 is bounded by the size of the initial derivation at the same depth.

Consider Π_1, that is the derivation obtained from Π_0 by reducing all the redexes at depth 1; by Theorem 1 we have $\mathsf{W}(\Pi_1, j) = \mathsf{W}(\Pi_0, j) < \mathsf{W}(\Pi, 0)$ for $j < 1$ and $\mathsf{W}(\Pi_1, 1) < (\mathsf{W}(\Pi, 1) \cdot \mathsf{R}(\Pi))^{n_0}$, while $\mathsf{W}(\Pi_1, j) \leq \mathsf{W}(\Pi_0, j) \cdot (\sum_{h=1}^{\mathsf{d}(\Pi_0)} \mathsf{R}(\Pi_0, h))^{n_1} \leq \mathsf{W}(\Pi, j) \cdot (\mathsf{R}(\Pi))^{n_0+n_1}$ for $j > 1$; moreover, $n_1 \leq \mathsf{W}(\Pi, 1) \cdot (\mathsf{R}(\Pi))^{n_0}$ since the number of reductions at depth 1 is bounded by the size of Π_0 at depth 1.

By repeating the same reasoning, if we reduce up to a generic depth d, then we obtain $\mathsf{W}(\Pi_d, j) \leq \mathsf{W}(\Pi, j) \cdot (\mathsf{R}(\Pi))^{\sum_{e=0}^{d-1} n_e}$ for every $j > d$, and moreover $n_d \leq (\mathsf{W}(\Pi, d) \cdot \sum_{h=d}^{\mathsf{d}(\Pi)} \mathsf{R}(\Pi, h))^{\sum_{e=0}^{d-1} n_e}$. It is easy to see that, for $d \geq 1$, we have:

- $\sum_{e=0}^{d} \mathsf{W}(\Pi_i, e) \leq \sum_{e=0}^{d-1} \mathsf{W}(\Pi_i, e) + \mathsf{W}(\Pi, d) \cdot (\mathsf{R}(\Pi))^{\sum_{e=0}^{d-1} n_e}$;
- $\sum_{e=0}^{d} n_e \leq \sum_{e=0}^{d-1} n_e + \mathsf{W}(\Pi, d) \cdot (\mathsf{R}(\Pi))^{\sum_{e=0}^{d-1} n_e}$.

Then the functions f_d, g_d are defined by induction on d as follows:

$$f_0(x, y) = x$$
$$f_{d+1}(x, y) = f_d(x, y) + \mathsf{W}(\Pi) \cdot (\mathsf{R}(\Pi))^{g_d(x,y)}$$

$$g_0(x, y) = x$$
$$g_{d+1}(x, y) = g_d(x, y) + \mathsf{W}(\Pi) \cdot (\mathsf{R}(\Pi))^{g_d(x,y)}$$

These functions can be easily shown to be elementary by induction on d.

The following theorem is a consequence of the previous result, obtained by considering a reduction sequence up to the maximum depth:

Theorem 2. *There are elementary functions* $\phi, \psi : \mathbb{N} \times \mathbb{N} \to \mathbb{N}$ *such that, for every* $\Pi \triangleright \Gamma \mid \Delta \vdash \mathsf{M} : \sigma$, *the lenght of reduction sequences starting from* M *is at most* $\phi(\mathsf{W}(\Pi), \mathsf{R}(\Pi))$ *and the lenght of every reduct is at most* $\psi(\mathsf{W}(\Pi), \mathsf{R}(\Pi))$.

Proof. The proof follows directly from Lemma 4 by considering $\phi = f_{\mathsf{d}(\Pi)}$ and $\psi = f_{\mathsf{d}(\Pi)}$, taking into account the fact that $\mathsf{d}(\Pi)$ does not increase during the reduction, and that, by definition, $\mathsf{R}(\Pi) = \sum_{j=0}^{\mathsf{d}(\Pi)} \mathsf{R}(\Pi, j)$.

5   Conclusion

We presented a type assignment system for the call-by-value λ-calculus, such that every typable term can be reduced to its normal form in a number of steps which depends in an elementary way on both the size of the term and of the rank of the derivation. Observe that every typable term can be assigned an infinite number of types, so every type derivation for it is a witness of its normalisation bound; however, for every typable term, we can consider a derivation of minimal rank typing it. Therefore, the bound depends only on the term, since the rank of the derivation depends only on it. .

The tool we used for reaching this result is inspired by intersection types; indeed, our multiset types and the equivalence relation defined on them can be seen as a non idempotent and non associative intersection. The lack of associativity supplies a notion of stratification, which recalls the stratification of the light logics, but is fairly less restrictive. In fact the system of this paper has a typability power stronger that the system ETAS of [7], where types are formulae of Elementary Linear Logic. The lack of idempotence supplies a notion of rank of a derivation, which is also a bound on the size of types that are assigned to any value occurring in it.

Note that the lack of idempotence would actually give a very easy proof of the strong normalisation property of typable terms and a bound on the number of normalisation steps. Indeed it is easy to check that $\Pi \rhd \Gamma \mid \Delta \vdash M : \sigma$ and $M \rightarrow_v M'$ imply $\Pi' \rhd \Gamma \mid \Delta \vdash M' : \sigma$, where $R(\Pi') < R(\Pi)$. An obvious consequence of this fact is that M is strongly normalising in a number of steps bounded by $R(\Pi)$. However such bound is much less interesting, since it is not expressed as a function of the size of M. In order to obtain the elementary bound we need to consider a more refined measure, which is the size of the term and the rank of the derivation at every depth. We are aware that the rank of the derivation as a bound for the types of values is an overestimation; indeed we are looking for a more realistic measure.

A natural question is whether system $\mathcal{M}$ is complete for elementary time, i.e., whether all the elementary functions can be coded in it. As of now, the system does not allow a uniform encoding of datatypes, such as natural numbers or binary words, because of the lack of idempotence; consider for example the natural number $n \geq 1$, which can be assigned in $\mathcal{M}$ the types (among others)

$$\underbrace{[a \rightarrow a, \ldots, a \rightarrow a]}_{n} \rightarrow [a] \rightarrow [a] \qquad \underbrace{[a \rightarrow a, \ldots, a \rightarrow a]}_{n} \rightarrow [a \rightarrow a]$$

Observe that, in both cases, the type of n depends on n itself; indeed in system $\mathcal{M}$ natural numbers do not share a common type. In principle this is not a problem, since we can define functions taking into account this kind of typing, but the all construction can become quite involved. We are working in this direction. In any case system $\mathcal{M}$ is an essential step towards a characterization of Elementary Time functions through intersection types.

A   Appendix

Proof of Lemma 3 *(Substitution Lemma).* The proof of Substitution Lemma uses the following lemma, which is a corollary of Lemma 2.

Lemma 5

i. Let $\Pi \triangleright \Gamma \mid \Delta \vdash M : \sigma$, where $M \in V$ and σ is a multiset type; then, for every $\sigma_1, \ldots, \sigma_n$ such that $\sigma \cong \sqcup_{i=1}^n \sigma_i$, there are $\Pi_i \triangleright \Gamma_i \mid \Delta_i \vdash M : \sigma_i$, for $1 \le i \le n$, such that $\Gamma = \sqcup_{i=1}^n \Gamma_i$ and $\Delta = \sqcup_{i=1}^n \Delta_i$; moreover, $W(\Pi, j) = \sum_{i=1}^n \frac{|\sigma_i|}{|\sigma|} \cdot W(\Pi_i, j)$ and $R(\Pi, j) = \sum_{i=1}^n R(\Pi_i, j)$ for every $j \ge 0$.

ii. Let $\Pi \triangleright \Gamma \mid \Delta \vdash M : \sigma$, where $M \in V$ and σ is a multiset type; then, for every $\sigma_1, \ldots, \sigma_n$ such that $\sigma \cong [\sigma_1, \ldots, \sigma_n]$, there are $\Pi_i \triangleright \Gamma_i \mid \Delta_i \vdash M : \sigma_i$, for $1 \le i \le n$, such that $\Delta = \sqcup_{i=1}^n \Gamma_i \sqcup_{i=1}^n [\Delta_i]$, $W(\Pi, j) = \sum_{i=1}^n \frac{|\sigma_i|}{|\sigma|} \cdot W(\Pi_i, j-1)$ and $R(\Pi, j) = \sum_{i=1}^n R(\Pi_i, j-1)$ for every $j > 0$.

Proof. The proof of point i can be given by induction on σ, that of point ii by induction on Π, using Lemma 2.

Now we are able to show the proof of Substitution Lemma. Both points follow by induction on Π.

i.   – Let Π be

$$\frac{}{x : [A_1] \mid \emptyset \vdash x : A_1} \ (Ax)$$

where $n = 1$, $M[N/x] = N$ and $\sigma = A_1$; then $\mathcal{S}_\Pi^{\Sigma_1}$ is $\Sigma_1 \triangleright \Gamma_1 \mid \Delta_1 \vdash N : A_1$, where $W(\mathcal{S}_\Pi^{\Sigma_1}, j) = W(\Sigma_1, j) \le W(\Pi, j) + W(\Sigma_1, j)$ and $R(\mathcal{S}_\Pi^{\Sigma_1}, j) = R(\Sigma_1, j) \le R(\Pi, j) + R(\Sigma_1, j)$ for every $j \ge 0$.

 – Let Π end with an application of rule (w). If Π is

$$\frac{\Pi' \triangleright \Gamma', x : [A_1, \ldots, A_n] \mid \Delta' \vdash M : \sigma}{\Gamma, x : [A_1, \ldots, A_n] \mid \Delta \vdash M : \sigma} \ (w)$$

then the proof follows by induction. Otherwise, let Π be

$$\frac{\Pi' \triangleright \Gamma', x : [A_1, \ldots, A_k] \mid \Delta' \vdash M : \sigma}{\Gamma, x : [A_1, \ldots, A_n] \mid \Delta \vdash M : \sigma} \ (w)$$

for some k such that $0 \le k < n$. By inductive hypothesis $\mathcal{S}_{\Pi'}^{\Sigma_1, \ldots, \Sigma_k} \triangleright \Gamma' \sqcup_{i=1}^k \Gamma_i \mid \Delta' \sqcup_{i=1}^k \Delta_i \vdash M[N/x] : \sigma$, where $W(\mathcal{S}_{\Pi'}^{\Sigma_1, \ldots, \Sigma_k}, j) \le W(\Pi', j) + \sum_{i=1}^k W(\Sigma_i, j)$ and $R(\mathcal{S}_{\Pi'}^{\Sigma_1, \ldots, \Sigma_k}, j) \le R(\Pi, j) + \sum_{i=1}^k R(\Sigma_i, j)$ for every $j \ge 0$. Then $\mathcal{S}_\Pi^{\Sigma_1, \ldots, \Sigma_n} \triangleright \Gamma \sqcup_{i=1}^n \Gamma_i \mid \Delta \sqcup_{i=1}^n \Delta_i \vdash M[N/x] : \sigma$ is obtained from $\mathcal{S}_{\Pi'}^{\Sigma_1, \ldots, \Sigma_k}$ by applying rule (w), where $W(\mathcal{S}_\Pi^{\Sigma_1, \ldots, \Sigma_n}, j) \le W(\Pi', j) + \sum_{i=1}^k W(\Sigma_i, j) \le W(\Pi, j) + \sum_{i=1}^n W(\Sigma_i, j)$ and $R(\mathcal{S}_\Pi^{\Sigma_1, \ldots, \Sigma_n}, j) = R(\mathcal{S}_{\Pi'}^{\Sigma_1, \ldots, \Sigma_k}, j) \le R(\Pi', j) + \sum_{i=1}^k R(\Sigma_i, j) \le R(\Pi, j) + \sum_{i=1}^n R(\Sigma_i, j)$ for every $j \ge 0$.

– Let Π be

$$\frac{\Pi' \triangleright \Gamma, \mathbf{x} : [A_1, \ldots, A_n], \mathbf{y} : [A] \mid \Delta \vdash P : \mu}{\Gamma, \mathbf{x} : [A_1, \ldots, A_n] \mid \Delta \vdash \lambda \mathbf{y}.P : A \to \mu} \ (\to I_L)$$

where $M = \lambda \mathbf{y}.P$ and $\sigma = A \to \mu$. By the α-rule we may assume that $\mathbf{y} \notin FV(N)$. By inductive hypothesis $\mathcal{S}_{\Pi'}^{\Sigma_1, \ldots, \Sigma_n} \triangleright \Gamma \sqcup_{i=1}^n \Gamma_i, \mathbf{y} : [A] \mid \Delta \sqcup_{i=1}^n \Delta_i \vdash P[N/\mathbf{x}] : \mu$, where $W(\mathcal{S}_{\Pi'}^{\Sigma_1, \ldots, \Sigma_n}, j) \leq W(\Pi', j) + \sum_{i=1}^n W(\Sigma_i, j)$ and $R(\mathcal{S}_{\Pi'}^{\Sigma_1, \ldots, \Sigma_n}, j) \leq R(\Pi', j) + \sum_{i=1}^n R(\Sigma_i, j)$ for every $j \geq 0$.

Then $\mathcal{S}_{\Pi}^{\Sigma_1, \ldots, \Sigma_n} \triangleright \Gamma \sqcup_{i=1}^n \Gamma_i \mid \Delta \sqcup_{i=1}^n \Delta_i \vdash (\lambda \mathbf{y}.P)[N/\mathbf{x}] : A \to \mu$ is obtained from $\mathcal{S}_{\Pi'}^{\Sigma_1, \ldots, \Sigma_n}$ by applying rule $(\to I_L)$, where $W(\mathcal{S}_{\Pi}^{\Sigma_1, \ldots, \Sigma_n}, j) \leq W(\Pi, j) + \sum_{i=1}^n W(\Sigma_i, j)$ by the definition of weight and $R(\mathcal{S}_{\Pi}^{\Sigma_1, \ldots, \Sigma_n}, j) = R(\mathcal{S}_{\Pi'}^{\Sigma_1, \ldots, \Sigma_n}, j) \leq R(\Pi, j) + \sum_{i=1}^n R(\Sigma_i, j)$ for every $j \geq 0$.
The case of $(\to I_M)$ is similar.

– Let Π end with an application of rule $(\to E)$ with premises $\Pi_1 \triangleright \Gamma', \mathbf{x} : [A_1, \ldots, A_k] \mid \Delta' \vdash P : \rho \to \sigma$ and $\Pi_2 \triangleright \Gamma'', \mathbf{x} : [A_{k+1}, \ldots, A_n] \mid \Delta'' \vdash Q : \rho$, such that $M = PQ$. By inductive hypothesis on Π_1, there is a derivation $\mathcal{S}_{\Pi_1}^{\Sigma_1, \ldots, \Sigma_k} \triangleright \Gamma' \sqcup_{i=1}^k \Gamma_i \mid \Delta' \sqcup_{i=1}^k \Delta_i \vdash P[N/\mathbf{x}] : \rho \to \sigma$, where $W(\mathcal{S}_{\Pi_1}^{\Sigma_1, \ldots, \Sigma_k}, j) \leq W(\Pi_1, j) + \sum_{i=1}^k W(\Sigma_i, j)$ and $R(\mathcal{S}_{\Pi_1}^{\Sigma_1, \ldots, \Sigma_k}, j) \leq R(\Pi_1, j) + \sum_{i=1}^k R(\Sigma_i, j)$ for every $j \geq 0$. By inductive hypothesis on Π_2, there is a derivation $\mathcal{S}_{\Pi_2}^{\Sigma_{k+1}, \ldots, \Sigma_n} \triangleright \Gamma'' \sqcup_{i=k+1}^n \Gamma_i \mid \Delta'' \sqcup_{i=k+1}^n \Delta_i \vdash Q[N/\mathbf{x}] : \rho$, where $W(\mathcal{S}_{\Pi_2}^{\Sigma_{k+1}, \ldots, \Sigma_n}, j) \leq W(\Pi_2, j) + \sum_{i=k+1}^n W(\Sigma_i, j)$ and $R(\mathcal{S}_{\Pi_2}^{\Sigma_{k+1}, \ldots, \Sigma_n}, j) \leq R(\Pi_2, j) + \sum_{i=k+1}^n R(\Sigma_i, j)$ for every $j \geq 0$. Then $\mathcal{S}_{\Pi}^{\Sigma_1, \ldots, \Sigma_n} \triangleright \Gamma' \sqcup \Gamma'' \sqcup_{i=1}^n \Gamma_i \mid \Delta' \sqcup \Delta'' \sqcup_{i=1}^n \Delta_i \vdash (PQ)[N/\mathbf{x}] : \sigma$ is obtained by applying rule $(\to E)$ to the premises $\mathcal{S}_{\Pi_1}^{\Sigma_1, \ldots, \Sigma_k}$ and $\mathcal{S}_{\Pi_2}^{\Sigma_{k+1}, \ldots, \Sigma_n}$, where $W(\mathcal{S}_{\Pi}^{\Sigma_1, \ldots, \Sigma_n}, j) \leq W(\Pi, j) + \sum_{i=1}^n W(\Sigma_i, j)$ follows by the definition of weight and $R(\mathcal{S}_{\Pi}^{\Sigma_1, \ldots, \Sigma_n}, j) \leq R(\Pi_1, j) + \sum_{i=1}^k R(\Sigma_i, j) + R(\Pi_2, j) + \sum_{i=k+1}^n R(\Sigma_i, j) \leq R(\Pi, j) + \sum_{i=1}^n R(\Sigma_i, j)$ for every $j \geq 0$.

– The case of (m) is not possible.

ii. – The case of (Ax) is not possible.

– Let Π end with an application of rule (w). If Π is

$$\frac{\Pi' \triangleright \Gamma' \mid \Delta', \mathbf{x} : \tau \vdash M : \sigma}{\Gamma \mid \Delta, \mathbf{x} : \tau \vdash M : \sigma} \ (w)$$

then the proof follows by induction. Otherwise, let Π be

$$\frac{\Pi' \triangleright \Gamma \mid \Delta, \mathbf{x} : \tau_1 \vdash M : \sigma}{\Gamma \mid \Delta, \mathbf{x} : \tau_1 \sqcup \tau_2 \vdash M : \sigma} \ (w)$$

where $\tau \cong \tau_1 \sqcup \tau_2$. Note that $|\tau| = |\tau_1| + |\tau_2|$. By Lemma 5.i there are $\Sigma_1 \triangleright \Gamma_1 \mid \Delta_1 \vdash N : \tau_1$ and $\Sigma_2 \triangleright \Gamma_2 \mid \Delta_2 \vdash N : \tau_2$ such that $\Gamma' = \Gamma_1 \sqcup \Gamma_2$ and $\Delta' = \Delta_1 \sqcup \Delta_2$; moreover, $W(\Sigma, j) = \sum_{i=1}^2 \frac{|\tau_i|}{|\tau|} \cdot W(\Sigma_i, j)$ and $R(\Sigma, j) = \sum_{i=1}^2 R(\Sigma_i, j)$ for every $j \geq 0$. By inductive hypothesis $\mathcal{S}_{\Pi'}^{\Sigma_1} \triangleright \Gamma_1 \sqcup \Gamma \mid \Delta_1 \sqcup \Delta \vdash M[N/\mathbf{x}] : \sigma$, where $W(\mathcal{S}_{\Pi'}^{\Sigma_1}, j) \leq W(\Pi', j) + |\tau_1| \cdot$

$W(\Sigma_1, j)$ and $R(\mathcal{S}_{\Pi'}^{\Sigma_1}, j) \leq R(\Pi', j) + R(\Sigma_1, j)$ for every $j \geq 0$.

Then $\mathcal{S}_{\Pi}^{\Sigma} \rhd \Gamma \sqcup \Gamma' \mid \Delta \sqcup \Delta' \vdash M[N/x] : \sigma$ is obtained from $\mathcal{S}_{\Pi'}^{\Sigma_1}$ by applying rule (w), where $W(\mathcal{S}_{\Pi}^{\Sigma}, j) = W(\mathcal{S}_{\Pi'}^{\Sigma_1}, j) = W(\Pi', j) + |\tau_1| \cdot W(\Sigma_1, j) \leq W(\Pi, j) + |\tau| \cdot W(\Sigma, j)$ and $R(\mathcal{S}_{\Pi}^{\Sigma}, j) = R(\mathcal{S}_{\Pi'}^{\Sigma_1}, j) \leq R(\Pi', j) + R(\Sigma_1, j) \leq R(\Pi, j) + R(\Sigma, j)$ for every $j \geq 0$.

– Let Π be

$$\frac{\Pi' \rhd \Gamma, \mathbf{y} : [A] \mid \Delta, \mathbf{x} : \tau \vdash P : \mu}{\Gamma \mid \Delta, \mathbf{x} : \tau \vdash \lambda \mathbf{y}.P : A \to \mu} \ (\to I_L)$$

where $M = \lambda \mathbf{y}.P$.

By the α-rule we may assume that $\mathbf{y} \notin FV(N)$. By inductive hypothesis $\mathcal{S}_{\Pi'}^{\Sigma} \rhd \Gamma \sqcup \Gamma', \mathbf{y} : [A] \mid \Delta \sqcup \Delta' \vdash P[N/x] : \mu$, where $W(\mathcal{S}_{\Pi'}^{\Sigma}, j) \leq W(\Pi', j) + |\tau| \cdot W(\Sigma, j)$ and $R(\mathcal{S}_{\Pi'}^{\Sigma}, j) \leq R(\Pi', j) + R(\Sigma, j)$ for every $j \geq 0$. Then $\mathcal{S}_{\Pi}^{\Sigma} \rhd \Gamma \sqcup \Gamma' \mid \Delta \sqcup \Delta' \vdash (\lambda \mathbf{y}.P)[N/x] : A \to \mu$ is obtained from $\mathcal{S}_{\Pi'}^{\Sigma}$ by applying rule $(\to I)$, where $W(\mathcal{S}_{\Pi}^{\Sigma}, j) \leq W(\Pi, j) + |\tau| \cdot W(\Sigma, j)$ follows by the definition of weight and $R(\mathcal{S}_{\Pi}^{\Sigma}, j) = R(\mathcal{S}_{\Pi'}^{\Sigma}, j) \leq R(\Pi', j) + R(\Sigma, j) = R(\Pi, j) + R(\Sigma, j)$ for every $j \geq 0$.

The case of $(\to I_M)$ is similar.

– Let Π end with an application of rule $(\to E)$ with premises $\Pi_1 \rhd \Gamma_1 \mid \Delta_1, \mathbf{x} : \tau_1 \vdash P : \rho \to \sigma$ and $\Pi_2 \rhd \Gamma_2 \mid \Delta_2, \mathbf{x} : \tau_2 \vdash Q : \rho$, where $\tau \cong \tau_1 \sqcup \tau_2$ and $M = PQ$. Note that $|\tau| = |\tau_1| + |\tau_2|$. By Lemma 5.i, $\Sigma \rhd \Gamma' \mid \Delta' \vdash N : \tau$ implies there are $\Sigma_1 \rhd \Gamma_1' \mid \Delta_1' \vdash N : \tau_1$ and $\Sigma_2 \rhd \Gamma_2' \mid \Delta_2' \vdash N : \tau_2$ such that $\Gamma' = \Gamma_1' \sqcup \Gamma_2'$ and $\Delta' = \Delta_1' \sqcup \Delta_2'$; moreover, $W(\Sigma, j) = \sum_{i=1}^{2} \frac{|\tau_i|}{|\tau|} \cdot W(\Sigma_i, j)$ and $R(\Sigma, j) = \sum_{i=1}^{2} W(\Sigma_i, j)$ for every $j \geq 0$. By inductive hypothesis on Π_1, there is a derivation $\mathcal{S}_{\Pi_1}^{\Sigma_1} \rhd \Gamma_1 \sqcup \Gamma_1' \mid \Delta_1 \sqcup \Delta_1' \vdash P[N/x] : \rho \to \sigma$, where $W(\mathcal{S}_{\Pi_1}^{\Sigma_1}, j) \leq W(\Pi_1, j) + |\tau_1| \cdot W(\Sigma_1, j)$ and $R(\mathcal{S}_{\Pi_1}^{\Sigma_1}, j) \leq R(\Pi_1, j) + R(\Sigma_1, j)$ for every $j \geq 0$. Similarly, by inductive hypothesis on Π_2 there is $\mathcal{S}_{\Pi_2}^{\Sigma_2} \rhd \Gamma_2 \sqcup \Gamma_2' \mid \Delta_2 \sqcup \Delta_2' \vdash Q[N/x] : \rho$, where $W(\mathcal{S}_{\Pi_2}^{\Sigma_2}, j) \leq W(\Pi_2, j) + |\tau_2| \cdot W(\Sigma_2, j)$ and $R(\mathcal{S}_{\Pi_2}^{\Sigma_2}, j) \leq R(\Pi_2, j) + R(\Sigma_2, j)$ for every $j \geq 0$.

Then $\mathcal{S}_{\Pi}^{\Sigma} \rhd \Gamma_1 \sqcup \Gamma_2 \sqcup \Gamma' \mid \Delta_1 \sqcup \Delta_2 \sqcup \Delta' \vdash (PQ)[N/x] : \sigma$ is obtained by applying rule $(\to E)$ to $\mathcal{S}_{\Pi_1}^{\Sigma_1}$ and $\mathcal{S}_{\Pi_2}^{\Sigma_2}$, where $W(\mathcal{S}_{\Pi}^{\Sigma}, j) \leq W(\Pi, j) + |\tau_1| \cdot W(\Sigma_1, j) + |\tau_2| \cdot W(\Sigma_2, j) = W(\Pi, j) + |\tau| \cdot W(\Sigma, j)$ and $R(\mathcal{S}_{\Pi}^{\Sigma}, j) = R(\mathcal{S}_{\Pi_1}^{\Sigma_1}, j) + R(\mathcal{S}_{\Pi_2}^{\Sigma_2}, j) \leq R(\Pi_1, j) + R(\Sigma_1, j) + R(\Pi_2, j) + R(\Sigma_2, j) = R(\Pi, j) + R(\Sigma, j)$ for every $j \geq 0$.

– Let Π end with an application of rule (m): there are two possible cases.

• Let Π be

$$\frac{\Pi_s \rhd \Gamma_s, \mathbf{x} : [A_s^1, \ldots, A_s^{h_s}] \mid \Delta_s \vdash M : \sigma_s \quad (1 \leq s \leq p)}{\emptyset \mid \sqcup_{s=1}^{p} \Gamma_s, \sqcup_{s=1}^{p} [\Delta_s], \mathbf{x} : \tau \vdash M : \sigma} \ (m)$$

where $\tau \cong [A_1^1, \ldots, A_1^{h_1}, \ldots, A_p^1, \ldots, A_p^{h_p}]$ and $\Delta = \sqcup_{s=1}^{p} \Gamma_s, \sqcup_{s=1}^{p} [\Delta_s]$. By hypothesis there is $\Sigma \rhd \Gamma' \mid \Delta' \vdash N : \tau$. By Lemma 5.ii, there are $\Sigma_s^r \rhd \Gamma_s^r \mid \Delta_s^r \vdash N : A_s^r$ $(1 \leq s \leq p, 1 \leq r \leq h_s)$ such that $\Delta' = \sqcup_{s=1}^{p} {}_{r=1}^{h_s} \Gamma_s^r, \sqcup_{s=1}^{p} {}_{r=1}^{h_s} [\Delta_s^r]$; moreover $W(\Sigma, 0) = R(\Sigma, 0) = 0$, $W(\Sigma, j + 1) = \sum_{s=1}^{p} {}_{r=1}^{h_s} \frac{1}{|\tau|} \cdot W(\Sigma_s^r, j)$ and $R(\Sigma, j + 1) = \sum_{s=1}^{p} {}_{r=1}^{h_s} R(\Sigma_s^r, j)$ for every $j \geq 0$. By Lemma 3.i, there are derivations $\mathcal{S}_{\Pi_s}^{\Sigma_s^1, \ldots, \Sigma_s^{h_s}} \triangleright \Gamma_s \sqcup_{r=1}^{h_s}$ $\Gamma_s^r \mid \Delta_s \sqcup_{r=1}^{h_s} \Delta_s^r \vdash \mathtt{M[N/x]} : \sigma_s$, where $\mathtt{W}(\mathcal{S}_{\Pi_s}^{\Sigma_s^1, \ldots, \Sigma_s^{h_s}}, j) \leq \mathtt{W}(\Pi_s, j) +$ $\sum_{r=1}^{h_s} \mathtt{W}(\Sigma_s^r, j)$ and $\mathtt{R}(\mathcal{S}_{\Pi_s}^{\Sigma_s^1, \ldots, \Sigma_s^{h_s}}, j) \leq \mathtt{R}(\Pi_s, j) + \sum_{r=1}^{h_s} \mathtt{R}(\Sigma_s^r, j)$ for every $j \geq 0$ and $1 \leq s \leq p$. Then $\mathcal{S}_{\Pi}^{\Sigma} \triangleright \Gamma' \mid \Delta \sqcup \Delta' \vdash \mathtt{M[N/x]} :$ $[\sigma_1, \ldots, \sigma_n]$ is obtained by applying rule (m) to $\mathcal{S}_{\Pi_s}^{\Sigma_s^1, \ldots, \Sigma_s^{h_s}}$, for $1 \leq$ $s \leq p$, followed by a (possibly empty) sequence of rule (w) adding Γ', where $\mathtt{W}(\mathcal{S}_{\Pi}^{\Sigma}, 0) = \mathtt{R}(\mathcal{S}_{\Pi}^{\Sigma}, 0) = 0$, $\mathtt{W}(\mathcal{S}_{\Pi}^{\Sigma}, j + 1) = \sum_{s=1}^{p} \frac{|\sigma_s|}{|\sigma|} \cdot$ $\mathtt{W}(\mathcal{S}_{\Pi_s}^{\Sigma_s^1, \ldots, \Sigma_s^{h_s}}, j) \leq \sum_{s=1}^{p} \frac{|\sigma_s|}{|\sigma|} \cdot (\mathtt{W}(\Pi_s, j) + \sum_{r=1}^{h_s} \mathtt{W}(\Sigma_s^r, j)) = \mathtt{W}(\Pi, j +$ $1) + \sum_{s=1}^{p} \frac{|\sigma_s|}{|\sigma|} \cdot \sum_{r=1}^{h_s} \mathtt{W}(\Sigma_s^r, j) \leq \mathtt{W}(\Pi, j + 1) + \sum_{s=1, r=1}^{p, h_s} \mathtt{W}(\Sigma_s^r, j) \leq$ $\mathtt{W}(\Pi, j + 1) + |\tau| \cdot \mathtt{W}(\Sigma, j + 1)$ and $\mathtt{R}(\mathcal{S}_{\Pi}^{\Sigma}, j + 1) = \sum_{s=1}^{p} \mathtt{R}(\mathcal{S}_{\Pi_s}^{\Sigma_s^1, \ldots, \Sigma_s^{h_s}}, j) \leq$ $\sum_{s=1}^{p} \mathtt{R}(\Pi_s, j) + \sum_{s=1, r=1}^{p, h_s} \mathtt{R}(\Sigma_s^r, j) = \mathtt{R}(\Pi, j + 1) + \mathtt{R}(\Sigma, j + 1)$ for every $j \geq 0$.

- Let Π be

$$\frac{\Pi_s \triangleright \Gamma_s \mid \Delta_s, \mathtt{x} : \tau_s \vdash \mathtt{M} : \sigma_s \quad (1 \leq s \leq p)}{\emptyset \mid \sqcup_{s=1}^{p} \Gamma_s, \sqcup_{s=1}^{p} [\Delta_s], \mathtt{x} : \tau \vdash \mathtt{M} : \sigma} \ (m)$$

where $\Delta = \sqcup_{s=1}^{p} \Gamma_s, \sqcup_{s=1}^{p} [\Delta_s]$ and $\tau \cong [\tau_1, \ldots, \tau_s]$.

By hypothesis there is $\Sigma \triangleright \Gamma' \mid \Delta' \vdash \mathtt{N} : \tau$. By Lemma 5.ii, there are $\Sigma_s \triangleright \Gamma_s' \mid \Delta_s' \vdash \mathtt{N} : \tau_s$ $(1 \leq s \leq p)$ such that $\Delta' = \sqcup_{s=1}^{p} \Gamma_s', \sqcup_{s=1}^{p} [\Delta_s']$; moreover $\mathtt{W}(\Sigma, 0) = \mathtt{R}(\Sigma, 0) = 0$, $\mathtt{W}(\Sigma, j+1) = \sum_{s=1}^{p} \frac{|\tau_s|}{|\tau|} \cdot \mathtt{W}(\Sigma_s, j)$ and $\mathtt{R}(\Sigma, j + 1) = \sum_{s=1}^{p} \mathtt{R}(\Sigma_s, j)$ for every $j \geq 0$.

By inductive hypothesis, there are $\mathcal{S}_{\Pi_s}^{\Sigma_s} \triangleright \Gamma_s \sqcup \Gamma_s' \mid \Delta_s \sqcup \Delta_s' \vdash$ $\mathtt{M[N/x]} : \sigma_s$, where $\mathtt{W}(\mathcal{S}_{\Pi_s}^{\Sigma_s}, j) \leq \mathtt{W}(\Pi_s, j) + |\tau_s| \cdot \mathtt{W}(\Sigma_s^r, j)$ and $\mathtt{R}(\mathcal{S}_{\Pi_s}^{\Sigma_s}, j) \leq$ $\mathtt{R}(\Pi_s, j) + \mathtt{R}(\Sigma_s, j)$ for every $j \geq 0$ and $1 \leq s \leq p$.

Then $\mathcal{S}_{\Pi}^{\Sigma} \triangleright \Gamma' \mid \Delta \sqcup \Delta' \vdash \mathtt{M[N/x]} : [\sigma_1, \ldots, \sigma_n]$ is obtained by applying rule (m) to $\mathcal{S}_{\Pi_1}^{\Sigma_1}, \ldots, \mathcal{S}_{\Pi_p}^{\Sigma_p}$, followed by a (possibly empty) sequence of applications of rule (w) adding Γ', where $\mathtt{W}(\mathcal{S}_{\Pi}^{\Sigma}, 0) = \mathtt{R}(\mathcal{S}_{\Pi}^{\Sigma}, 0) = 0$, $\mathtt{W}(\mathcal{S}_{\Pi}^{\Sigma}, j + 1) = \sum_{s=1}^{p} \frac{|\sigma_s|}{|\sigma|} \cdot \mathtt{W}(\mathcal{S}_{\Pi_s}^{\Sigma_s}, j) \leq \sum_{s=1}^{p} \frac{|\sigma_s|}{|\sigma|} \cdot (\mathtt{W}(\Pi_s, j) + |\tau_s| \cdot \mathtt{W}(\Sigma_s, j)) = \mathtt{W}(\Pi, j + 1) + \sum_{s=1}^{p} \frac{|\sigma_s|}{|\sigma|} \cdot |\tau_s| \cdot \mathtt{W}(\Sigma_s, j) \leq \mathtt{W}(\Pi, j + 1) + \sum_{s=1}^{p} |\tau_s| \cdot \mathtt{W}(\Sigma_s, j) = \mathtt{W}(\Pi, j+1) + |\tau| \cdot \mathtt{W}(\Sigma, j + 1)$ by Lemma 5.ii, and $\mathtt{R}(\mathcal{S}_{\Pi}^{\Sigma}, j + 1) = \sum_{s=1}^{p} \mathtt{R}(\mathcal{S}_{\Pi_s}^{\Sigma_s}, j) \leq \sum_{s=1}^{p} \mathtt{R}(\Pi_s, j) + \sum_{s=1}^{p} \mathtt{R}(\Sigma_s, j) = \mathtt{R}(\Pi, j + 1) + \mathtt{R}(\Sigma, j + 1)$ for every $j \geq 0$.

References

1. Asperti, A., Roversi, L.: Intuitionistic light affine logic. ACM Trans. Comput. Logic 3(1), 137–175 (2002)
2. Baillot, P., De Benedetti, E., Ronchi Della Rocca, S.: Characterizing polynomial and exponential complexity classes in elementary lambda-calculus. In: Diaz, J., Lanese, I., Sangiorgi, D. (eds.) TCS 2014. LNCS, vol. 8705, pp. 151–163. Springer, Heidelberg (2014). doi:10.1007/978-3-662-44602-7_13

3. Baillot, P., Terui, K.: Light types for polynomial time computation in lambda-calculus. In: LICS 2004, pp. 266–275. IEEE Computer Society (2004)
4. Bernadet, A., Lengrand, S.: Non-idempotent intersection types and strong normalisation. Logical Methods Comput. Sci. **9**(4), 1–46 (2013)
5. de Carvalho, D.: Execution Time of lambda-Terms via Denotational Semantics and Intersectio Types. MSCS (2009). To appear
6. Coppo, M., Dezani-Ciancaglini, M.: An extension of the basic functionality theory for the lambda-calculus. Notre-Dame J. Formal Logic **21**(4), 685–693 (1980)
7. Coppola, P., Dal Lago, U., Ronchi Della Rocca, S.: Light logics and the call-by-value lambda calculus. Logical Methods Comput. Sci. **4**(4) (2008)
8. Danos, V., Joinet, J.: Linear logic and elementary time. Inf. Comput. **183**(1), 123–137 (2003)
9. De Benedetti, E., Ronchi Della Rocca, S.: Bounding normalization time through intersection types. In: ITRS 2012, vol. 121, pp. 48–57 (2013)
10. De Benedetti, E.: Ronchi Della Rocca, S.: A type assignment for lambda-calculus complete both for FPTIME and strong normalization. Inf. Comput. **248**, 195–214 (2016)
11. Gaboardi, M., Ronchi Della Rocca, S.: A soft type assignment system for λ-calculus. In: Duparc, J., Henzinger, T.A. (eds.) CSL 2007. LNCS, vol. 4646, pp. 253–267. Springer, Heidelberg (2007). doi:10.1007/978-3-540-74915-8_21
12. Girard, J.: Light linear logic. Inf. Comput. **143**(2), 175–204 (1998)
13. Lafont, Y.: Soft linear logic and polynomial time. Theor. Comput. Sci. **318**(1–2), 163–180 (2004)
14. Paolini, L., Piccolo, M., Ronchi Della Rocca, S.: Logical semantics for stability. In: MFPS 2009, vol. 249, pp. 429–449. Elsevier (2009)
15. Plotkin, G.D.: Call-by-name, call-by-value and the lambda-calculus. Theor. Comput. Sci. **1**(2), 125–159 (1975)
16. Ronchi Della Rocca, S., Paolini, L.: The Parametric λ-Calculus: A Metamodel for Computation. Texts in Theoretical Computer Science: An EATCS Series. Springer, Berlin (2004)

Probabilistic Resource Analysis by Program Transformation

Maja H. Kirkeby$^{(\boxtimes)}$ and Mads Rosendahl

Computer Science, Roskilde University, Roskilde, Denmark
{majaht,madsr}@ruc.dk

Abstract. The aim of a probabilistic resource analysis is to derive a probability distribution of possible resource usage for a program from a probability distribution of its input. We present an automated multi-phase rewriting based method to analyze programs written in a subset of C. It generates a probability distribution of the resource usage as a possibly uncomputable expression and then transforms it into a closed form expression using over-approximations. We present the technique, outline the implementation and show results from experiments with the system.

1 Introduction

The main contribution in this paper is to present a technique for probabilistic resource analysis where the analysis is seen as a program-to-program translation. This means that the transformation to closed form is a source code program transformation problem and not specific to the analysis. Any necessary approximations in the analysis are performed at the source code level. The technique also makes it possible to balance the precision of the analysis against the brevity of the result.

Many optimizations for increased energy efficiency require probabilistic and average case analysis as part of the transformations. Wierman et al. states that *"global energy consumption is affected by the average case, rather than the worst case"* [37]. Also in scheduling *"an accurate measurement of a task's average-case execution time can assist in the calculation of more appropriate deadlines"* [17]. For a subset of programs a precise average-case execution time can be found using static analysis [12,14,31]. Applications of such analysis may be in improving scheduling of operations or in temperature management. Because the analysis returns a distribution, it can be used to calculate the probability of energy consumptions above a certain limit, and thereby indicate the risk of over-heating.

The central idea in this paper is to use probabilistic output analysis in combination with a preprocessing phase that instruments programs with resource

M. Rosendahl—The research leading to these results has received funding from the European Union Seventh Framework Programme (FP7/2007-2013) under grant agreement no 318337, ENTRA - Whole-Systems Energy Transparency.

M. van Eekelen and U. Dal Lago (Eds.): FOPARA 2015, LNCS 9964, pp. 60–80, 2016.
DOI: 10.1007/978-3-319-46559-3_4

usage. We translate programs into an intermediate language program that computes the probability distribution of resource usage. This program is then analyzed, transformed, and approximated with the aim of obtaining a closed form expression. It is an alternative to deriving cost relations directly from the program [7] or expressing costs as abstract values in a semantics for the language.

As with automatic complexity analysis, the aim of probabilistic resource analysis is to express the result as a parameterized expression. The time complexity of a program should be expressed as a closed form expression in the input size, and for probabilistic resource analysis, the aim is to express the probability of resource usage of the program parameterized by input size or range. If input values are not independent, we can specify a joint distribution for the values. Values do not have to be restricted to a finite range but for infinite ranges the distribution would converge to zero towards the limit.

The current work extends our previous work on probabilistic analysis [29] in three ways. We show how to use a preprocessing phase to instrument programs with resource usage such that the resource analysis can be expressed as an analysis of the possible output of a program. The resource analysis can handle an extended class of programs with structured data as long as the program flow does not depend on the probabilistic data in composite data structures. Finally, we present an implementation of the analysis in the Ciao language [5] which uses algebraic reductions in the Mathematica system [39].

The focus in this paper is on using fairly simple local resource measures where we count core operations on data. Since the instrumentation is done at the source code level, we can use flow information so that the local costs can depend on actual data to operations and which operations are executed before and after. This is normally not relevant for time complexity but does play an important role for energy consumption analysis [19,32].

2   Probability Distributions in Static Analysis

In our approach to probabilistic analysis, the result of an analysis is an approximation of a probability distribution. We will here present the concepts and notation we will use in the rest of the paper. A probability distribution is also often referred to as the *probability mass function* in the discrete case, and in the continuous case, it is a *probability density functions*. We will use an upper case P letter to denote a probability distribution.

Definition 1 (input probability). *For a countable set X an input probability distribution is a mapping $P_X : X \to \{r \in \mathbb{R} \mid 0 \le r \le 1\}$, where*

$$\sum_{x \in X} P_X(x) = 1$$

We define the output probability distribution for a program p in a forward manner. It is the *weight* or sum of all probabilities of input values where the program returns the desired value z as output.

Definition 2 (output probability). *Given a program,* $p : X \to Z$ *and a probability distribution for the input,* P_X, *the output probability distribution,* $P_p(z)$, *is defined as:*

$$P_p(z) = \sum_{x \in X \land p(x) = z} P_X(x)$$

Note that Kozen also uses a similar forward definition [20], whereas Monniaux constructs the inverse mapping from output to input for each program statement and expresses the relationship in a backwards style [23].

Lemma 1. *The output probability distribution,* $P_p(z)$, *satisfies*

$$0 \leq \sum_z P_p(z) \leq 1$$

The program may not terminate for all input, and this means that the sum may be less than one. If we expand the domain Z with an element to denote non-termination, $Z_\perp$, the total sum of the output distribution $P_p(z)$ would be 1.

In our static analysis, we will use approximations to obtain safe and simplified results. Various approaches to approximations of probability distributions have been proposed and can be interpreted as *imprecise probabilities* [1,9,10]. Dempster-Shafer structures [2,16] and P-boxes [11] can be used to capture and propagate uncertainties of probability distributions. There are several results on extending arithmetic operations to probability distributions for both known dependencies between random variables and when the dependency is unknown or only partially known [3,4,18,33,38]. Algorithms for lifting basic operations on numbers to basic operations on probability distributions can be used as abstractions in static analysis based on abstract interpretation. Our approach uses the P-boxes as bounds of probability distributions. P-boxes are normally expressed in terms of the cumulative probability distribution but we will here use the probability mass function. We do not, however, use the various basic operations on P-boxes, but apply approximations to a probability program such that it forms a P-box.

Definition 3 (over-approximation). *For a distribution* P_p *an over-approximation* $(\overline{P}_p)$ *of the distribution satisfies the condition:*

$$\overline{P}_p : \forall z. P_p(z) \leq \overline{P}_p(z) \leq 1.$$

The aim of the probabilistic resource analysis is to derive an approximation $\overline{P}_p$ as tight as possible.

The over-approximation of the probability distribution can be used to derive lower and upper bounds of the expected value and will thus approximate the expected value as an interval [29].

3   Architecture of the Transformation System

The system contains five main phases. The input to the system is a program in a small subset of C with annotations of which part we want to analyze. It could be the whole program but can also be a specific subroutine which is called repeatedly with varying arguments according to some input distribution.

The first phase will instrument the program with resource measuring operations. The instrumented program will perform the same operations as the original program in addition to recording and printing resource usage information. This program can still be compiled and run, and it will also produce the same results as the original program.

The second phase translates the program into an intermediate language for further analysis. We use a small first-order functional language for the analysis process. The translation has two core elements. We slice [36] the program with respect to the resource measuring operations and transform loops into primitive recursion in the intermediate language. The transformed program can still be executed and will produce the same resource usage information as the instrumented program. Since the instrumentation is done before the translation to intermediate language any interpretation overhead or speed-up due to slicing does not influence the result [28].

In the third phase, we construct a probability output program that computes the probability output function. In this case, it is a probability distribution of possible resource usages of the original program. This program can also run but will often be extremely inefficient since it will merge information for all possible input to the original program.

The fourth phase transforms the probability program into a large expression without further function calls. Recursive calls are removed using summations and the transformed program computes the same result as the program did before this phase.

In the final phase, the probability function is transformed into closed form using symbolic summation and over-approximation. In this phase we exploit the Mathematica system [39]. The final probability program computes the same result or an over-approximation of the function produced in the fourth phase.

4   Instrumenting Programs for Resource Analysis

The input to the analysis is a program in a subset of C. In the next section we define the intermediate language for further analysis and it is the restrictions on the intermediate language that limits the source programs we can analyze with our system. The source program may contain integer variable and arrays, usual loop constructs and non-recursive function calls. The program should be annotated with specification on which part of the program to analyze. The following is an example of such a program.

```
// ToAnalyse: multa(_,_,_,N)
void multa(int a1[MX],int a2[MX],int a3[MX],int n){
  int i1,i2,i3,d;
  for(i1 = 0; i1 < n; i1++) {
    for(i2 = 0; i2 < n; i2++) {
      d = 0;
      for(i3 = 0; i3 < n; i3++) {
        d = d + a1[i1*n+i3]*a2[i3*n+i2];
      }
      a3[i1*n+i2] = d;
    }
  }
}
```

This example program describes a matrix multiplication for which we would like to analyze the probability distribution for the number of steps when parameterized with the size (N) of the matrices.

Instrumentation. The program is then instrumented with resource usage information and translated into an intermediate language for further analysis. The instrumented program is also a valid program in the source language and can be executed with the same results as the original program. It will, however, also collect resource usage information.

In our example, we instrument the program with step counting information where we count the number of assignment statement being executed. This is done by inserting a variable into the program and incrementing it once for each assignment statement.

```
int multa(int a1[MX],int a2[MX],int a3[MX],int n){
  int i1,i2,i3,d;
  int step; step=0;
  for(i1 = 0; i1 < n; i1++) {
    for(i2 = 0; i2 < n; i2++) {
      d = 0; step++;
      for(i3 = 0; i3 < n; i3++) {
        d = d + a1[i1*n+i3]*a2[i3*n+i2]; step++;
      }
      a3[i1*n+i2] = d; step++;
    }
  }
  return step;
}
```

The outer loop does not update the step counter, whereas the first inner loop updates it twice per iteration and the innermost loop updates it once per loop iteration.

Slicing. The second phase will slice the program with respect to resource usage and translate the program into the intermediate language of first order functions that we will use in the subsequent stages. Loops in the program are translated into primitive recursion.

```
for3(i3, step, n) =
  if(i3 = n) then step else for3(i3 + 1,step+1,n)

for2(i2, step, n) =
  if(i2 = n) then step else for2(i2 + 1,for3(0,step+2,n),n)

for1(i1, step, n) =
  if(i1 = n) then step else for1(i1 + 1,for2(0,step,n),n)

tmulta(n)= for1(0,step,n)
```

Each function in the recursive program corresponds to a for loop with their related step-updates. The step counter is given as input argument to the next function in a continuation-passing style.

Intermediate Language. An intermediate program, Prg, consists of integer functions, $f_i\colon Int^* \to Int$, as given by the abstract syntax given in Fig. 1. In the following, we relax the restrictions on function and parameter names.

$$f_i(x_1, \ldots, x_n) \stackrel{\text{def}}{=} \langle\exp\rangle$$
$$\langle\text{aexp}\rangle \models x_i \mid c \mid \langle\text{aexp}\rangle +_i \langle\text{aexp}\rangle \mid \langle\text{aexp}\rangle -_i \langle\text{aexp}\rangle \mid$$
$$\langle\text{aexp}\rangle \times_i \langle\text{aexp}\rangle \mid \langle\text{aexp}\rangle \, \text{div}_i \, \langle\text{aexp}\rangle$$
$$\langle\text{bexp}\rangle \models \langle\text{aexp}\rangle =_i \langle\text{aexp}\rangle \mid \langle\text{aexp}\rangle <_i \langle\text{aexp}\rangle \mid \langle\text{aexp}\rangle \leq_i \langle\text{aexp}\rangle \mid$$
$$\text{true} \mid \text{false} \mid \text{not}(\langle\text{bexp}\rangle)$$
$$\langle\exp\rangle \models \langle\text{aexp}\rangle \mid f_i(\langle\exp_1\rangle, \ldots, \langle\exp_n\rangle) \mid$$
$$\text{if } \langle\text{bexp}\rangle \text{ then } \langle\exp\rangle \text{ else } \langle\exp\rangle$$

Fig. 1. The abstract syntax describing the intermediate programs.

Definition 4. *A program is* well-formed *if it follows the abstract syntax and it contains a finite number of function definitions, that each is of one of the following forms and can internally be enumerated with a natural number such that:*

$f_i(x_1, \ldots, x_n) \stackrel{\text{def}}{=} \text{if } b \text{ then } e_0 \text{ else } f_i(e_1, \ldots, e_n)$
where f_i is simple, e_0 only contains calls to functions f_j where $j < i$.
$f_i(x_1, \ldots, x_n) \stackrel{\text{def}}{=} e$
where e only contain calls to functions f_j where $j < i$.

The enumeration prevents mutual recursion and ensures that non-recursive calls cannot create an infinite call-chain.

5   Probabilistic Output Analysis

The analysis is applied to the intermediate program and an input probability program in the intermediate language. The output is a new program that can be described by a subset of the intermediate language; this will be clarified later in the definition of pure and closed form programs. The analysis consists of three phases:

Create, where the probability program describing the output distribution is created as a possibly uncomputable expression.

Separate, where we remove all calls from the probability program.

Simplify, where we transform the program into closed form using safe over-approximations when necessary.

The analysis is constructed as three sets of transformation rules, one for each of the three phases. All transformations are syntax directed, and a strategy is to apply them in a depth-first manner. The program output analysis is implemented in Ciao and integrates with Mathematica in the third phase to reduce expressions.

In the following we use $Var(e)$ to represent the set of variables occurring in expression e, and $f(x_1, ..., x_n) \overset{\text{def}}{=} e$ to represent the function f is defined in the input program. Some side conditions are explained in an informal way, as in "$f(x_1, ..., x_n) \overset{\text{def}}{=} e$, where e is non-recursive".

$$\text{name} \frac{\text{precondition}_1 \quad ... \quad \text{precondition}_n}{\text{original term} \rightarrow \text{rewritten term}}$$

The preconditions are evaluated from left to right, and if all succeeds, we can use the transformation. When substituting a variable x to an expression e, we denote it $[x/e]$.

In the following we will begin by extending the intermediate language presented in Fig. 1 such that it can express probabilities, and afterwards describe the transformation rules for each phase.

The Intermediate Language. The intermediate language is, as previously mentioned, a first-order functional language. A probability program can be evaluated at any stage through the transformation process.

We extend the abstract syntax given in Fig. 1 such that it can easily describe probability distributions. We introduce probability functions, P: $Int^* \rightarrow Real$, which follows the expanded syntax given in Fig. 2. The dots indicate the syntax described in Fig. 1. Again, ⟨aexp⟩ and ⟨exp⟩ are of type integer, ⟨bexp⟩ is boolean, and the new ⟨qexp⟩ is a real. In ⟨qexp⟩ the method i2r type casts an integer expression to a real. We introduce c, sum, prod and argDev functions. c evaluates to either 1, if its boolean expression evaluates to true, or 0 when it evaluates to false. Evaluating sum instantiates the variable with all possible values and sum all the results of the evaluation the ⟨qexp⟩. prod instantiates its variable with all values for which the first $\langle$qexp$\rangle$ evaluates to 1, and then it multiply all the results from evaluating the second $\langle$qexp$\rangle$. The last expression introduced is argDev which describes the development of the variable x_i as a function of the number of updates, x_j. The expression $\langle$exp$\rangle$ computes the development of x_i for one incrementation of x_j (e.g. the argument x_i in a function $f(x_i)$ with a recursive call $f(x_i -_i 2)$ has a argument development argDev(x_i, x_i-2, x_j)).

$$
\begin{aligned}
&f_i(x_1, \ldots, x_n) \overset{\text{def}}{=} \langle\text{exp}\rangle \\
&\qquad \langle\text{aexp}\rangle \models \ldots \mid \min(\langle\text{aexp}\rangle, \langle\text{aexp}\rangle) \mid \max(\langle\text{aexp}\rangle, \langle\text{aexp}\rangle) \\
&\qquad \langle\text{bexp}\rangle \models \ldots \mid \langle\text{aexp}\rangle =_i \langle\text{exp}\rangle \\
&\qquad \langle\text{exp}\rangle \models \ldots \mid \text{argDev}(x_i, \langle\text{exp}\rangle, x_j) \\
&P_i(x_1, \ldots, x_n) \overset{\text{def}}{=} \langle\text{qexp}\rangle \\
&\qquad \langle\text{qexp}\rangle \models \text{i2r}(\langle\text{aexp}\rangle) \mid \text{c}(\langle\text{bexp}\rangle) \mid \langle\text{qexp}\rangle \; op_q \; \langle\text{qexp}\rangle \mid \\
&\qquad\qquad \text{sum}(x_i, \langle\text{qexp}\rangle) \mid \text{prod}(x_i, \langle\text{qexp}\rangle, \langle\text{qexp}\rangle) \mid \\
&\qquad\qquad P_i(\langle\text{aexp}_1\rangle, \ldots, \langle\text{aexp}_n\rangle) \\
&op_q = +_q \mid -_q \mid \times_q \mid /^q
\end{aligned}
$$

Fig. 2. The expanded abstract syntax describing probability programs.

A program that computes a probability distribution is referred to as a probability program.

Definition 5. *A probability program that has no if-expressions no function calls is* pure *and a pure probability program without any* sum *and* prod *is in* closed form.

A program is *pure* after it is transformed in the separation phase and is pure and in *closed form* after the simplification phase.

The Create Phase. This phase has only one rule which creates a program that computes a probability distribution from the intermediate program and input distributions.

$$
\text{create} \frac{f(u_1, \ldots, u_n) \overset{\text{def}}{=} e \qquad P(v_1, \ldots, v_n) \overset{\text{def}}{=} e_p}{P_f(z) \overset{\text{def}}{=} \text{sum}(x_1 \, ; \ldots \text{sum}(x_n \, ; \text{c}(z =_i f(x_1, \ldots, x_n)) \times_q P(x_1, \ldots, x_n)))}
$$

We use the create rule to make a new probability function describing the probability distribution for the integer function we are interested in.

The Separate Phase. In this phase function calls are removed by repeatedly exposing calls and replacing them. Non-recursive function calls are unfolded using their definitions. Function calls can occur inside if-expressions or as nested calls; these are extracted and handled one at a time.

$$\text{f-simple}\,\frac{\mathtt{f}(\mathtt{y}_1,...,\mathtt{y}_n) \stackrel{\text{def}}{=} e \quad, \text{ where } e \text{ is non-recursive} \quad \mathtt{x}_1,...,\mathtt{x}_n \in \mathcal{V}ar}{\mathtt{c}(\mathtt{z} =_\mathtt{i} \mathtt{f}(\mathtt{x}_1,...,\mathtt{x}_n)) \to \mathtt{c}(\mathtt{z} =_\mathtt{i} e[\mathtt{y}_1/\mathtt{x}_1,...,\mathtt{y}_n/\mathtt{x}_n])}$$

$$\text{rem-P}\,\frac{\mathtt{P}(x_1,...,x_n) \stackrel{\text{def}}{=} e}{\mathtt{P}(e_1,...,e_n) \to e[x_1/e_1,...,x_n/e_n]}$$

$$\text{rem-if}\,\frac{}{\mathtt{c}(\mathtt{z} =_\mathtt{i} \text{if } b \text{ then } e_0 \text{ else } e_1) \to (\mathtt{c}(b) \times_\mathtt{q} \mathtt{c}(\mathtt{z} =_\mathtt{i} e_0) +_\mathtt{q} \mathtt{c}(\mathtt{not}(b)) \times_\mathtt{q} \mathtt{c}(\mathtt{z} =_\mathtt{i} e_1))}$$

$$\text{no-nest(f)}\,\frac{\{e_1,...,e_n\} \nsubseteq \mathcal{V}ar}{\begin{array}{l}\mathtt{c}(\mathtt{z} =_\mathtt{i} \mathtt{f}(e_1,...,e_n)) \to \\ \mathtt{sum}(\mathtt{u}_1 \,;...\mathtt{sum}(\mathtt{u}_n \,; \mathtt{c}(\mathtt{z} =_\mathtt{i} \mathtt{f}(\mathtt{u}_1,...,\mathtt{u}_n)) \times_\mathtt{q} \mathtt{c}(\mathtt{u}_1 =_\mathtt{i} e_1) \times_\mathtt{q}... \times_\mathtt{q} \mathtt{c}(\mathtt{u}_n =_\mathtt{i} e_n)))\end{array}}$$

We replace calls to recursive functions by a summation over the number of recursions using argument development constructors to describe the value of each argument as a function of the index of the summation. This way of defining argument development has similarities with size change functions derived using recurrence equations. Argument development functions do not depend on the base-case unlike size-change functions [40]. The summation also contains a product which ensures that the condition evaluates to false for argument values less than the current value of the index of summation. When the expression in a product contains only c-constructors, then the product is evaluated to 1 if either the range is empty or the expression is evaluated to true for the full range. The following rewrite rules are all that is needed for transforming probability programs into pure probability programs.

$$\text{f-rec}\,\frac{\mathtt{f}(\mathtt{y}_1,...,\mathtt{y}_n) \stackrel{\text{def}}{=} \text{if } b \text{ then } e_0 \text{ else } \mathtt{f}(e_1,...,e_n) \qquad \mathtt{x}_1,...\mathtt{x}_n \in \mathcal{V}ars}{\quad\sigma_{y/i} = [\mathtt{y}_1/\mathtt{i}_1,...,\mathtt{y}_n/\mathtt{i}_n]\sigma_{y/x} = [\mathtt{y}_1/\mathtt{x}_1,...,\mathtt{y}_n/\mathtt{x}_n]\sigma_{y/j} = [\mathtt{y}_1/\mathtt{j}_1,...,\mathtt{y}_n/\mathtt{j}_n]}$$
$$\begin{array}{l}\mathtt{c}(\mathtt{z} =_\mathtt{i} \mathtt{f}(\mathtt{x}_1,...,\mathtt{x}_n)) \to \\ \mathtt{sum}(\mathtt{i} \,; \mathtt{c}(0 \leq_\mathtt{i} \mathtt{i}) \times_\mathtt{q} \\ \quad \mathtt{sum}(\mathtt{i}_1 \,;...\mathtt{sum}(\mathtt{i}_n \,; \mathtt{c}(\sigma_{y/i}(b)) \times_\mathtt{q} \mathtt{c}(\mathtt{i}_1 =_\mathtt{i} \mathtt{argDev}(\mathtt{x}_1, \sigma_{y/x}(e_1), \mathtt{i})) \times_\mathtt{q} \\ \qquad \mathtt{c}(\mathtt{z} =_\mathtt{i} \sigma_{y/i}(e_0)) \times_\mathtt{q}... \times_\mathtt{q} \mathtt{c}(\mathtt{i}_n =_\mathtt{i} \mathtt{argDev}(\mathtt{x}_n, \sigma_{y/x}(e_n), \mathtt{i}))) ...) \times_\mathtt{q} \\ \quad \mathtt{prod}(\mathtt{j} \,; \mathtt{c}(0 \leq_\mathtt{i} \mathtt{j}) \times_\mathtt{q} \mathtt{c}(\mathtt{j} \leq_\mathtt{i} \mathtt{i} -_\mathtt{i} 1) \,; \\ \quad \mathtt{sum}(\mathtt{j}_1 \,;...\mathtt{sum}(\mathtt{j}_n \,; \mathtt{c}(\mathtt{not}(\sigma_{y/j}(b))) \times_\mathtt{q} \\ \qquad \mathtt{c}(\mathtt{j}_1 =_\mathtt{i} \mathtt{argDev}(\mathtt{x}_1, \sigma_{y/x}(e_1), \mathtt{j})) \times_\mathtt{q}... \times_\mathtt{q} \mathtt{c}(\mathtt{j}_n =_\mathtt{i} \mathtt{argDev}(\mathtt{x}_n, \sigma_{y/x}(e_n), \mathtt{j})) \\ \quad)...)))\end{array}$$

The argument development expression may contain function calls as well, and these are extracted equivalently to nested functions.

$$\text{no-nest(argDev)}\,\frac{}{\begin{array}{l}\mathtt{c}(\mathtt{z} =_\mathtt{i} \mathtt{argDev}(\mathtt{x}, \mathtt{f}(e_1,...,e_n), \mathtt{i})) \to \\ \mathtt{sum}(\mathtt{u} \,; \mathtt{c}(\mathtt{z} =_\mathtt{i} \mathtt{argDev}(\mathtt{x}, \mathtt{f}(e_1,...,e_n), \mathtt{i})) \times_\mathtt{q} \mathtt{c}(\mathtt{u} =_\mathtt{i} \mathtt{f}(e_1,...,e_n)))\end{array}}$$

After applying these rules until they cannot be applied any more, the probability program has been transformed to pure form.

The Simplification Phase. We have presented the rules for obtaining a pure probability program, and in this section we outline the rules used to reach closed form. A pure probability function consists of a series of nested summations multiplied with an expression (e.g. input probability). The rules are applied in no particular order and the phase ends when no more rules can be applied. In this phase we integrate with Mathematica. A call to Mathematica is denoted $\text{mm:Function}(Arg) = Answer$, where Function denotes the actual function called in Mathematica (e.g. mm:Expand calls Mathematica's Expand function). The translation between the intermediate language and Mathematica's representation will not be discussed further here.

The rules can be grouped by their functionality: preparing expressions, removal of summations and removal of products. The latter are currently the only rules containing over-approximations.

Preparing expressions for removal of either summations or products involve moving expressions that do not depend on the index of summation outside the summation, dividing summations of additions into simpler ones, reducing expressions, dividing summations in ranges, and remove argument development constructors. Please notice that $\text{div-sum}(x\leq)$ has an equivalent rule for upper bounds.

$$\text{move-c}\ \frac{x \notin \mathcal{V}ar(e_1)}{\text{sum}(x\,;e_1\times_q e_2) \rightarrow e_1\times_q\text{sum}(x\,;e_2)}$$

$$\text{div-sum}(+)\ \frac{x \in \mathcal{V}ar(e_1)\qquad x \in \mathcal{V}ar(e_2)}{\text{sum}(x\,;e_1 +_q e_2) \rightarrow \text{sum}(x\,;e_1) +_q \text{sum}(x\,;e_2)}$$

$$\text{div-sum}(x\leq)\ \frac{x \notin \mathcal{V}ar(e_1,c_2)\qquad x \in \mathcal{V}ar(e_2)}{\begin{array}{c}\text{sum}(x\,;c(x\leq_i e_1)\times_q c(x\leq_i e_2)\times_q e_3) \rightarrow \\ c(e_1 \leq_i e_2)\times_q\text{sum}(x\,;c(x\leq_i e_1)\times_q e_3) +_q \\ c(e_2 \leq_i e_1 -_i 1)\times_q\text{sum}(x\,;c(x\leq_i e_2)\times_q e_3)\end{array}}$$

$$\text{rem}(\text{argDev})\ \frac{c \in n}{c(z =_i \text{argDev}(x, x +_i c, i)) \rightarrow c(z =_i x +_i c\times_i i)}$$

$$\text{reduceAexp}\ \frac{\text{mm:Reduce}(e_1) = e_2}{c(e_1) \rightarrow c(e_2)}$$

$$\text{reduce}(=)\ \frac{}{c(\text{true}) \rightarrow \text{i2r}(1)}$$

Removal of summations can be done in two ways. Either the index of the summation can only be one value or it can be a limited range of values, and depending on which case different transformations are used. In the first case, there exists an equation containing the variable index of the innermost summation. The equation is solved for the variable, and the rest of the variable occurrences are replaced by the new value.

$$\text{rem-sum}(=)\ \frac{\text{mm:Solve}(e_1 =_i e_2, x) = c(x =_i e_3)}{\text{sum}(x\,;c(e_1 =_i e_2)\times_q e) \rightarrow e[x/e_3]}$$

Removing a summation by its range involves using standard mathematical formulas for rewriting series. The last part of the following rule uses $\sum_{k=1}^{n} k^2 = n(n+1)(2n+1)/6$. We only present transformations up to quadratic series and our pragmatic implementation contains rules for transforming series of power of degree up to 10. A more general rewrite rule for series of power of degree up p could be implemented, but is more complicated as it includes Bernoulli numbers and binomial coefficients. The precondition uses Mathematica's Expand to transform the expression into the right pattern.

$$\text{rem-sum}(\leq)\frac{x \notin Var(e_1, ..., e_6) \quad \text{mm:Expand}(e_3) = \text{i2r}(e_4 +_i e_5 \times_i x +_i e_6 \times_i x \times_i x)}{\begin{array}{l} \text{sum}(x\,;\,c(e_1 \leq_i x) \times_q c(x \leq_i e_2) \times_q \text{i2r}(e_3)) \rightarrow \\ \quad \text{i2r}(e_4) \times_q \text{i2r}(e_2 -_i e_1 +_i 1) +_q \\ \quad \text{i2r}(e_5) \times_q \text{i2r}(e_2 \times_i (e_2 +_i 1))/^q 2 -_q \\ \quad \text{i2r}(e_5) \times_q \text{i2r}(e_2 \times_i (e_2 -_i 1))/^q 2 +_q \\ \quad \text{i2r}(e_6) \times_q \text{i2r}(e_2 \times_i (e_2 +_i 1) \times_i (2 \times_i e_2 +_i 1))/^q 6 -_q \\ \quad \text{i2r}(e_6) \times_q \text{i2r}(e_2 \times_i (e_2 -_i 1) \times_i (2 \times_i e_2 -_i 1))/^q 6 \end{array}}$$

Removal of product involves a safe over-approximation. The implementation of POA contains two different over-approximations and in many cases the probability program can be transformed into closed form in a precise manner. In the following paragraph we describe when the transformation preserves the accuracy of the transformed term.

The probability function can always be over-approximated to 1. The rule f-rec is an exact rule and introduces a product-expression which may not be possible to rewrite into closed form. We only introduce the product-expression with c-expressions in its body, and therefore it will always either evaluate to 1 or to 0. A safe over-approximation of such a product-expression is 1.

$$\text{rem-prod-one}\frac{x \notin Var(e_1, e_2) \quad x \in Var(e_3)}{\text{prod}(x\,;\,c(e_1 \leq_i x) \times_q c(x \leq_i e_2)\,;\,c(e_3)) \rightarrow 1}$$

For the summation describing recursive calls, this transformation is exact when the condition, b, evaluates to true for exactly one value (eg. it is an equation).

A broader class of recursive programs (than those having an equation in the condition) is those where the c-expression is monotone in x; meaning that there exists a k for which $c(e_3) = 1$ for $x \leq k$ and $c(e_3) = 0$ for $x > k$. This case covers many for-loops. In this case, we can accurately replace the prod-expression with two c-expressions one checking the lower and one checking the upper range-limit. The empty product (the lower limit is larger than the upper) is 1.

$$\text{rem-prod-mon}\frac{x \notin Var(e_1, e_2) \quad x \in Var(e_3) \quad e_3 \text{ is monotone in } x}{\begin{array}{l} \text{prod}(x\,;\,c(e_1 \leq_i x) \times_q c(x \leq_i e_2)\,;\,c(e_3)) \rightarrow \\ (c(e_3)[x/e_1] \times_q c(e_3)[x/e_2] \times_q c(e_1 \leq_i e_2) +_q c(e_2 \leq_i e_1 -_i 1)) \end{array}}$$

This rule does not preserve accuracy when the c-expression is not monotone in x (e.g. $c(2 \leq_i x || 4 \leq_i x)$).

6 Implementation and Results

In the following, we present three examples which show results of programs with nested loops parameterized input distribution of multiple variables. The probability distribution computed by the output program varies in complexity; the first program calculates a single parameterized output, the second program computes a triangular shaped output distribution and third computes a distribution converging towards a standard normal distribution. The results are presented in a reduced and readable form extracted from our implementation.

Matrix Multiplication. The original matrix multiplication program uses composite types and contains nested loops. The intermediate program, defined in Fig. 3, contains nested recursive calls but has no dependency on data in composite types.

```
for3(i3,step,n) = if(i3>=n) then step else for3(i3+1,step+1,n)
for2(i2,step,n) = if(i2>=n) then step else for2(i2+1,for3(0,step+2,n),n)
for1(i1,step,n) = if(i1>=n) then step else for1(i1+1,for2(0,step,n),n)
tmulta(step,n) = for1(0,step,n)
P(step,n1) = c(step=0)*c(n1=n)
```

Fig. 3. The intermediate program containing also the parameterized probability distribution. The parameter n can obtain only one value.

The nested calls create argument development functions that depend on function calls. These are transformed into a simple form and then removed. The introduced products are over-approximated, but due to the form of the condition the result is precise. The output program computes a single value distribution (when specialized with the size of the matrix). It is given in Fig. 4 along with an array describing a subset of specializations of the output program with respect to a value of n.

	n	program
Ptmulta(out) =	1	Ptmulta(out) = c(out=3)
c(3=<out/(n*n))*	2	Ptmulta(out) = c(out=16)
c(1=<n)*	3	Ptmulta(out) = c(out=45)
c(out/n*n=2+n)*1	4	Ptmulta(out) = c(out=96)
	...	...

Fig. 4. The general output probability program (left) and the program specialized with the value of n (right).

Adding Parameterized Distributions. This example is a recursive program computing the addition of two numbers; the input program and the input probability distribution can be seen in Fig. 5. The output depends on both increasing and decreasing values. In this example, we use a parameter n as the upper limit of a range of input values. The input distribution describes two independent variables, each having a uniform distribution from 1 to n.

```
add(x,y) = if x=<0 then y else add(x-1,y+1)
P(x) = c(1=<x)*c(x=<n)*1/n
Pxy(x,y) = P(x)*P(y)
```

Fig. 5. The intermediate program containing both the function add and the input probability distribution. Here, the parameter n is used to describe a range.

The analysis gives a precise probability distribution and computes a triangular distribution (or pyramid shaped distribution). The output probability program is shown in Fig. 6 along with a graph depicting the pyramid shaped output probability distributions for different initializations of n. The lower bound on out arises from the input probability distribution and not from the condition. The upper bound 2*n of the analysis result shows that the output depends on both input variables, despite that one is increasing and the other is decreasing.

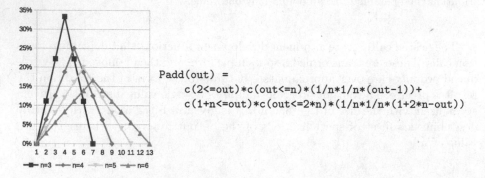

```
Padd(out) =
     c(2<=out)*c(out<=n)*(1/n*1/n*(out-1))+
     c(1+n<=out)*c(out<=2*n)*(1/n*1/n*(1+2*n-out))
```

Fig. 6. The general output program and the graphs for the output probability distribution with n set to 3, 4, 5, and 6, respectively.

Adding 4 Independent Variables. The program sum4 adds four variables and was presented by Monniaux [23]. Certain over-approximations were applied so as to obtain a safe and simplified result.

The program is recursive and in this example we use independent input variables each uniformly distributed input from 1 to 6, as described in Fig. 7.

Despite the ranges and their associated value are not symmetric, the resulting program computes a precise and perfectly symmetric probability distribution as

```
add(x,y) = if x=0 then y else add(x-1,y+1)
sum4(x,y,z,w) = add(x,add(y,add(z,w)))
tsum4(x,y,z,w) = sum4(x,y,z,w)
P(x) = c(1=<x)*c(x=<6)*1/6
Pxyzw(x,y,z,w) = P(x)*P(y)*P(z)*P(w)
```

Fig. 7. Intermediate program.

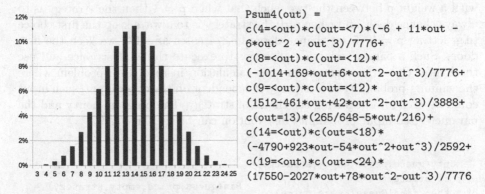

```
Psum4(out) =
c(4=<out)*c(out=<7)*(-6 + 11*out -
6*out^2 + out^3)/7776+
c(8=<out)*c(out=<12)*
(-1014+169*out+6*out^2-out^3)/7776+
c(9=<out)*c(out=<12)*
(1512-461*out+42*out^2-out^3)/3888+
c(out=13)*(265/648-5*out/216)+
c(14=<out)*c(out=<18)*
(-4790+923*out-54*out^2+out^3)/2592+
c(19=<out)*c(out=<24)*
(17550-2027*out+78*out^2-out^3)/7776
```

Fig. 8. The output program and graph for its computed probability distribution for out from 3 to 25.

shown in Fig. 8. The difference in the ranges comes (among other things) from the range dividing rules, as they do not divide the range symmetrically. As expected from the central limit theorem of probability theory, the resulting probability program describes a distribution that has similarities with a normal distribution.

Monty Hall. The Monty Hall problem is often used to exemplify how gained knowledge influences probabilities (conditional probability). In this problem there are three closed doors; one hiding a price and two that are empty. The doors have an equal chance of hiding the price. There is a contestant, who should choose one of the doors, then the game host will open an empty door and the contestant can either stick with the first choice or can change to the other unopened door. The problem lies in showing whether the best winning-strategy is to stick with the first choice or to switch to the other?

If the strategy is to stick with the first choice and that door has a price then the contestant has won. If the contestant changes door he/she only loses if the first choice was the door hiding the price; if the first choice was an empty door, then the game host would open the other empty door leaving only the price door for a second choice.

The program monty models the two strategies; if the strategy variable is 1 then the strategy is to change the door, and otherwise the strategy is to stick with the first choice. The program takes as input the contestant's first guess, the door hiding the price, the empty door which is not opened by the game host and the strategy the contestant uses.

Let us assume the contestant has an equal chance of choosing each of the doors. The input variables guess, price, and empty models the first choice, the price door and the empty door which is left after the game host has opened an empty door. All three doors have a value between 1 and 3, and the empty door cannot be the same as the price door. We have parameterized the strategy with a weight p between the two, such that when p = 1 then the strategy is to always change door, and when p=0 the strategy is to always keep the first choice (e.g. letting p = 0.75 we change doors in 3/4 cases and 1/4 we keep the first door). Such a parameterization allows us to execute the analysis once and use the lighter closed form result for that calculation instead. In a problem where the winning-probability of a strategy is dependent on the other input, such input could be used for optimizing the choice of strategy. The program monty and the parameterized input probability distribution can be seen in Fig. 9.

```
monty(guess,price,empty,strategy)=
   if strategy = 0
   then finalGuess(guess,price)
   else change(guess,price,empty)

finalGuess(guess,price)=
   if price=guess then 1 else 0

change(guess,price,empty)=
   if price=guess
   then finalGuess(empty,price)
   else finalGuess(price,price)
```

```
Pin(guess,price,empty,strategy) =
   1/18*c(1=<guess)*c(guess=<3)
      *c(1=<price)*c(price=<3)
      *c(1=<empty)*c(empty=<3)
      *c(not(price = empty))
      *Pstrat(strategy)
Pstrat(strategy) =
   p*c(1 = strategy)
      + (1-p)*c(0 = strategy)
```

Fig. 9. The program monty models the event flow depending on the chosen strategy; if the strategy is 0 then the contestant keeps the first door and if it is 1 then the contestant changes his mind. There are three doors and the input of monty describes the contestants first guess, the door hiding the price, the empty door which is not opened by the game host (and is different from the price door) and the strategy of the contestant. If the final choice hides the price then the program returns 1 and otherwise 0. The probability of the strategy is an expression parameterized with a weight, p between the two strategies instead of executing the analysis twice.

The analysis was capable of handling the program correctly and the result can be seen in Fig. 10.

The probabilities 1/3 and 2/3 does not occur directly in the output probability program but are found in the constants 6, 12 and 1/18.

Adding Dependent Non-uniform Variables. A function call may have interdependent and non-uniform arguments that can only be represented as a

```
pmonty(out) =
  1/18 *
  (c(out=0)*
    (12*(1-p)+6*p)+
   c(out=1)*
    (6*(1-p)+12*p))
```

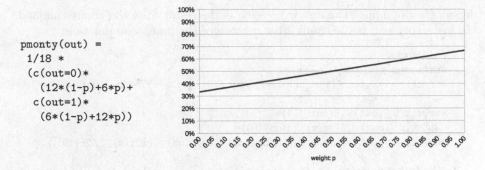

Fig. 10. The probability of winning the Monty Hall as a function of the weight given to change-strategy. The probabilistic output analysis reveals that the best weight between the keep strategy and the change strategy is to always use change strategy.

joint probability distribution (i.e. polygon), and in this example, we demonstrate that the analysis can handle such function calls. We focus on the dependencies, analyze a simple add program and discuss the limits of the interdependencies. The program also shows that interdependencies quickly lead to the occurrence of integer division in the output.

The input arguments are interdependent; the second argument is always less than or equal to the value of the first argument and the joint distribution depends only on the value of the first argument resulting in a triangular and skewed probability distribution. The probability program is defined in Fig. 11.

```
Pxy(x,y) =  c(1=<y)*c(y=<3) *
                c(1=<x)*c(x=<y) * x/10
add(x,y) = x+z

Padd(out) =
  c(2 =< out)*c(out=< 3) *   1/20 * out%2 * (1 + out%2) +
  c(4 =< out)*c(out=< 6) * -(1/20)*(-4+out-out%2)*(-3+out+out%2)
```

Fig. 11. An input program, add, its skewed joint distribution, Pxy, and the closed form probability program, Padd, produced by the analysis. The integer division is noted by a "%".

The create rule generates nested summations, and removing such inner summations imply that their values must be expressed using the variables of the outer summations or the input variable (i.e. out). Comparing the result from this experiment with the output probability distribution for addition of two random variables in Fig. 6 indicates that integer division is a special case arising from dependent input. The following interesting expressions are extracted during analysis execution, and they show how the integer division arises from the dependency of input. The first expression is the result from the create rule and the last expression is the result after removal of the inner y-summation.

$$P_{add}(\mathtt{out}) =$$
$$\mathtt{sum}(\mathtt{x}\,;\,\mathtt{sum}(\mathtt{y}\,;\,\mathtt{c}(\mathtt{out} =_i \mathtt{x} +_i \mathtt{y}) \times_q$$
$$\quad \mathtt{c}(1 \leq_i \mathtt{x}) \times_q \mathtt{c}(\mathtt{x} \leq_i \mathtt{y}) \times_q \mathtt{c}(1 \leq_i \mathtt{y}) \times_q \mathtt{c}(\mathtt{y} \leq_i 3) \times_q (\mathtt{i2r}(\mathtt{x}) /^q \mathtt{i2r}(10)))) =$$

$$\mathtt{sum}(\mathtt{x}\,;\,\mathtt{c}(2 \leq_i \mathtt{out}) \times_q \mathtt{c}(\mathtt{out} \leq_i 3) \times_q \mathtt{c}(1 \leq_i \mathtt{x}) \times_q$$
$$\quad \mathtt{c}(2 \times_i \mathtt{x} \leq_i \mathtt{out}) \times_q (\mathtt{i2r}(\mathtt{x}) /^q \mathtt{i2r}(10))) +_q$$
$$\mathtt{sum}(\mathtt{x}\,;\,\mathtt{c}(4 \leq_i \mathtt{out}) \times_q \mathtt{c}(\mathtt{out} \leq_i 3 +_i \mathtt{x}) \times_q \mathtt{c}(2 \times_i \mathtt{x} \leq_i \mathtt{out}) \times_q (\mathtt{i2r}(\mathtt{x}) /^q \mathtt{i2r}(10)))$$

In the last expression, there are two summations, each leading to its own part in the resulting program. Looking closely at each summation, we see that they share the upper limit for $\mathtt{x}$, $\mathtt{c}(2 \times_i \mathtt{x} \leq_i \mathtt{out})$, which currently contains an integer multiplication and when solved with respect to $\mathtt{x}$ contains the integer division. In the final result, the second part of the expression has an upper limit for $\mathtt{out}$, $\mathtt{c}(\mathtt{out} \leq_i 6)$ which is a constraint that the summation-removal-rule introduces to ensure that the lower limit of the summation (i.e. $\mathtt{out} -_i 3$) is less than or equal to the upper limit (i.e. $\mathtt{out} \%_i 2$).

The original probability $(\mathtt{i2r}(\mathtt{x}) /^q \mathtt{i2r}(10))$ occurs directly in the summations, and this indicates a limit of this implementation and approach. To be able to handle a probability, the rewrite rules for summations must transform summations over the probability expression. There are limits to which series the system currently can transform, Sum of reciprocals (e.g. $\sum_{k=1}^{n} \frac{1}{k}$) known as harmonic series or variations thereof such as generalized harmonic series are currently not implemented. The current analysis is limited to finite summations of at least order of 1, but a closer integration with Mathematica that exploits more of Mathematicas rewriting mechanisms should be able to handle such series.

7   Related Works

Probabilistic analysis is related to the analysis of probabilistic programs. Probabilistic analysis analysize programs with a normal semantics where the input variables are interpreted over probability distributions. Analysis of probabilistic programs analyze programs with probabilistic semantics where the values of the input variables are unknown (e.g. flow analysis [25]).

In probabilistic analysis, it is important to determine how variables depend on each other, but already in 1976 Denning proposed a flow analysis for revealing whether variables depend on each other [8]. This was presented in the field of secure flow analysis. Denning introduced a lattice-based analysis where she, given the name of a variable, that should be kept secret, deducted which other variables those should be kept secret in order to avoid leaking information. In 1996, Denning's method was refined by Volpano et al. into a type system and for the first time, it was proven sound [34].

Reasoning about probabilistic semantics is a closely related area to probabilistic analysis, as they both work with nested probabilistic influence. The probabilistic analysis work on standard semantic and analyze it using input probability distributions, where a probabilistic semantics allow for random assignments and probabilistic choices [20] and is normally analyzed using an expanded classical analysis or verification method [6].

Probabilistic model checking is an automated technique for formally verifying quantitative properties for systems with probabilistic behaviors. It is mainly focused on Markov decision processes, which can model both stochastic and non-deterministic behavior [13, 21]. It differs from probabilistic analysis as it assumes the Markov property.

In 2000, Monniaux applied abstract interpretation to programs with probabilistic semantics and gained safe bounds for worst-case analysis [23]. Pierro et al. introduce a linear mapping structure, a Moore-Penrose pseudo-inversè, instead of a Galois connection. They use the linear structures to compare 'closeness' of approximations as an expression using the average approximation error. Pierro et al. further explores using probabilistic abstract interpretation to calculate the average case analysis [24]. In 2012, Cousot and Monerau gave a general probabilistic abstraction framework [6] and stated, in Sect. 5.3, that Pierro et al.'s method and many other abstraction methods can be expressed in this new framework.

When analyzing probabilities the main challenge is to maintain the dependencies throughout the program. Schellekens defines this as *Randomness preservation* [31] (or random bag preservation) which in his (and Gao's [14]) case enables tracking of certain data structures and their distributions. They use special data structures as they find these suitable to derive the average number of basic operations. In another approach [26, 35], tests in programs has been assumed to be independent of previous history, also known as the Markov property (the probability of true is fixed). As Wegbreit remarked, this is true only for some programs (e.g. linear search for repeating lists) and others, this is not the case (linear search for non-repeating lists). The Markov property is the foundation in Markov decision processes which is used in probabilistic model-checking [13]. Cousot et al. present a probabilistic abstraction framework where they divide the program semantics into probabilistic behavior and (non-)deterministic behavior. They propose handling of tests when it is possible to assume the Markov property, and handle loops by using a probability distribution describing the probability of entering the loop in the ith iteration. Monniaux proposes another approach for abstracting probabilistic semantics [23]; he first lifts a normal semantics to a probabilistic semantics where random generators are allowed and then uses an abstraction to reach a closed form. Monniaux's semantic approach uses a backward probabilistic semantics operating on measurable functions. This is closely related to the forward probabilistic semantics proposed earlier by Kozen [20].

An alternative approach to probabilistic analysis is based on symbolic execution of programs with symbolic values [15]. Such techniques can also be used on programs with infinitely many execution paths by limiting the analysis to a finite set of paths at the expense of tightness of probability intervals [30].

8   Conclusion

Probabilistic analysis of program has a renewed interest for analyzing programs for energy consumptions. Numerous embedded systems and mobile applications are limited by restricted battery life on the hardware. In this paper, we describe a rewrite system that derives a resource probability distribution for programs given distributions of the input. It can analyze programs in subset of C where we have known distribution of input variables. From the original program we create a probability distribution program, where we remove calls to original functions and transform it into closed form. We have presented the transformation rules for each step and outlined the implementation of the system. We discuss over-approximating rules and their influence on the accuracy of the output probability and show that our analysis improves on related analysis in the literature.

References

1. Adje, A., Bouissou, O., Goubault-Larrecq, J., Goubault, E., Putot, S.: Static analysis of programs with imprecise probabilistic inputs. In: Cohen, E., Rybalchenko, A. (eds.) VSTTE 2013. LNCS, vol. 8164, pp. 22–47. Springer, Heidelberg (2014). doi:10.1007/978-3-642-54108-7_2
2. Bauer, M.: Approximations for decision making in the Dempster-Shafer theory of evidence. In: Horvitz, E., Jensen, F.V. (eds.) UAI, pp. 73–80. Morgan Kaufmann (1996)
3. Berleant, D., Cheng, H.: A software tool for automatically verified operations on intervals and probability distributions. Reliable Comput. 4(1), 71–82 (1998)
4. Bouissou, O., Goubault, E., Goubault-Larrecq, J., Putot, S.: A generalization of p-boxes to affine arithmetic. Computing 94(2–4), 189–201 (2012)
5. Bueno, F., Cabeza, D., Carro, M., Hermenegildo, M., López-Garcıa, P., Puebla, G.: The ciao prolog system. Reference Manual. The Ciao System Documentation Series-TR CLIP3/97.1, School of Computer Science, Technical University of Madrid (UPM), vol. 95, p. 96 (1997)
6. Cousot, P., Monerau, M.: Probabilistic abstract interpretation. In: Seidl, H. (ed.) ESOP 2012. LNCS, vol. 7211, pp. 169–193. Springer, Heidelberg (2012). doi:10.1007/978-3-642-28869-2_9
7. Debray, S.K., García, P.L., Hermenegildo, M., Lin, N.-W.: Estimating the computational cost of logic programs. In: Charlier, B. (ed.) SAS 1994. LNCS, vol. 864, pp. 255–265. Springer, Heidelberg (1994). doi:10.1007/3-540-58485-4_45
8. Denning, D.E.: A lattice model of secure information flow. Commun. ACM 19(5), 236–243 (1976)
9. Destercke, S., Dubois, D.: The role of generalised p-boxes in imprecise probability models. In: 6th International Symposium on Imprecise Probability: Theories and Applications (2009)
10. Ferson, S.: Model uncertainty in risk analysis. Technical report, Centre de Recherches de Royallieu, Universite de Technologie de Compiegne (2014)
11. Ferson, S., Kreinovich, V., Ginzburg, L., Myers, D.S., Sentz, K.: Constructing probability boxes and Dempster-Shafer structures. Sand 2002–4015, Sandia National Laboratories (2002)

12. Flajolet, P., Salvy, B., Zimmermann, P.: Automatic average-case analysis of algorithm. Theor. Comput. Sci. **79**(1), 37–109 (1991)
13. Forejt, V., Kwiatkowska, M., Norman, G., Parker, D.: Automated verification techniques for probabilistic systems. In: Bernardo, M., Issarny, V. (eds.) SFM 2011. LNCS, vol. 6659, pp. 53–113. Springer, Heidelberg (2011). doi:10.1007/978-3-642-21455-4_3
14. Gao, A.: Modular average case analysis: Language implementation and extension. Ph.d. thesis, University College Cork (2013)
15. Geldenhuys, J., Dwyer, M.B., Visser, W.: Probabilistic symbolic execution. In: Proceedings of the 2012 International Symposium on Software Testing and Analysis, pp. 166–176. ACM (2012)
16. Gordon, J., Shortliffe, E.H.: The Dempster-Shafer theory of evidence. In: Rule-Based Expert Systems: The MYCIN Experiments of the Stanford Heuristic Programming Project, p. 21 (1984)
17. Guo, X., Boubekeur, M., McEnery, J., Hickey, D.: ACET based scheduling of soft real-time systems: an approach to optimise resource budgeting. Int. J. Comput. Commun. **1**(1), 82–86 (2007)
18. Kay, R.U.: Fundamentals of the Dempster-Shafer theory and its applications to system safety and reliability modelling. In: RTA, pp. 173–185 (2007)
19. Kerrison, S., Eder, K.: Energy modelling and optimisation of software for a hardware multi-threaded embedded microprocessor. University of Bristol, Bristol, Technical report (2013)
20. Kozen, D.: Semantics of probabilistic programs. J. Comput. Syst. Sci. **22**(3), 328–350 (1981)
21. Kwiatkowska, M., Norman, G., Parker, D.: Advances and challenges of probabilistic model checking. In: 48th Annual Allerton Conference on Communication, Control, and Computing, pp. 1691–1698. IEEE, September 2010
22. Liqat, U., Kerrison, S., Serrano, A., Georgiou, K., Lopez-Garcia, P., Grech, N., Hermenegildo, M.V., Eder, K.: Energy consumption analysis of programs based on XMOS ISA level models. In: Gupta, G., Peña, R. (eds.) LOPSTR 2013. LNCS, vol. 8901, pp. 72–90. Springer, Heidelberg (2013). doi:10.1007/978-3-319-14125-1_5
23. Monniaux, D.: Abstract interpretation of probabilistic semantics. In: Palsberg, J. (ed.) SAS 2000. LNCS, vol. 1824, pp. 322–339. Springer, Heidelberg (2000). doi:10.1007/978-3-540-45099-3_17
24. Pierro, A., Hankin, C., Wiklicky, H.: Abstract interpretation for worst and average case analysis. In: Reps, T., Sagiv, M., Bauer, J. (eds.) Program Analysis and Compilation, Theory and Practice. LNCS, vol. 4444, pp. 160–174. Springer, Heidelberg (2007). doi:10.1007/978-3-540-71322-7_8
25. Di Pierro, A., Wiklicky, H., Puppis, G., Villa, T.: Probabilistic data flow analysis: a linear equational approach. In: Proceedings Fourth International Symposium, vol. 119, pp. 150–165. Open Publishing Association (2013)
26. Soza Pollman, H., Carro, M., Lopez Garcia, P.: Probabilistic cost analysis of logic programs: a first case study. INGENIARE - Revista Chilena de Ingeniera **17**(2), 195–204 (2009)
27. Rosendahl, M.: Automatic program analysis. Master's thesis, University of Copenhagen (1986)
28. Rosendahl, M.: Automatic complexity analysis. In: FPCA, pp. 144–156 (1989)
29. Rosendahl, M., Kirkeby, M.H.: Probabilistic output analysis by program manipulation. In: Quantitative Aspects of Programming Languages, EPTCS (2015)

30. Sankaranarayanan, S., Chakarov, A., Gulwani, S.: Static analysis for probabilistic programs: inferring whole program properties from finitely many paths. In: Proceedings of the 34th ACM SIGPLAN Conference on Programming Language Design and Implementation, pp. 447–458. ACM, June 2013

31. Schellekens, M.P.: A Modular Calculus for the Average Cost of Data Structuring. Springer, New York (2008)

32. Tiwari, V., Malik, S., Wolfe, A.: Power analysis of embedded software a first step towards software power minimization. IEEE Trans. Very Large Scale Integr. (VLSI) Syst. $2(4)$, 437–445 (1994)

33. Uwimbabazi, A.: Extended probabilistic symbolic execution. Master's thesis, University of Stellenbosch (2013)

34. Volpano, D.M., Irvine, C.E., Smith, G.: A sound type system for secure flow analysis. J. Comput. Secur. $4(2/3)$, 167–188 (1996)

35. Wegbreit, B.: Mechanical program analysis. Commun. ACM $18(9)$, 528–539 (1975)

36. Weiser, M.: Program slicing. In: Proceedings of the 5th International Conference on Software Engineering, pp. 439–449. IEEE Press (1981)

37. Wierman, A., Andrew, L.L.H., Tang, A.: Stochastic analysis of power-aware scheduling. In: Proceedings of Allerton Conference on Communication, Control and Computing. Urbana-Champaign, IL (2008)

38. Wilson, N.: Algorithms for Dempster-Shafer theory. In: Kohlas, J., Moral, S. (eds.) Handbook of Defeasible Reasoning and Uncertainty Management Systems. Handbook of Defeasible Reasoning and Uncertainty Management Systems, vol. 5, pp. 421–475. Springer, Netherlands (2000)

39. Wolfram, S.: The Mathematica Book. Cambridge University Press and Wolfram Research Inc., New York (2000)

40. Zuleger, F., Gulwani, S., Sinn, M., Veith, H.: Bound analysis of imperative programs with the size-change abstraction. In: Yahav, E. (ed.) SAS 2011. LNCS, vol. 6887, pp. 280–297. Springer, Heidelberg (2011). doi:10.1007/978-3-642-23702-7_22

Inferring Parametric Energy Consumption Functions at Different Software Levels: ISA vs. LLVM IR

U. Liqat[1], K. Georgiou[2], S. Kerrison[2], P. Lopez-Garcia[1,3(✉)],
John P. Gallagher[5], M.V. Hermenegildo[1,4], and K. Eder[2]

[1] IMDEA Software Institute, Madrid, Spain
{umer.liqat,pedro.lopez,manuel.hermenegildo}@imdea.org
[2] University of Bristol, Bristol, UK
{kyriakos.georgiou,steve.kerrison,kerstin.eder}@bristol.ac.uk
[3] Spanish Council for Scientific Research (CSIC), Madrid, Spain
[4] Universidad Politécnica de Madrid (UPM), Madrid, Spain
[5] Roskilde University, Roskilde, Denmark
jpg@ruc.dk

Abstract. The static estimation of the energy consumed by program executions is an important challenge, which has applications in program optimization and verification, and is instrumental in energy-aware software development. Our objective is to estimate such energy consumption in the form of *functions on the input data sizes of programs*. We have developed a tool for experimentation with static analysis which infers such energy functions at two levels, the instruction set architecture (ISA) and the intermediate code (LLVM IR) levels, and reflects it upwards to the higher source code level. This required the development of a translation from LLVM IR to an intermediate representation and its integration with existing components, a translation from ISA to the same representation, a resource analyzer, an ISA-level energy model, and a mapping from this model to LLVM IR. The approach has been applied to programs written in the XC language running on XCore architectures, but is general enough to be applied to other languages. Experimental results show that our LLVM IR level analysis is reasonably accurate (less than 6.4 % average error vs. hardware measurements) and more powerful than analysis at the ISA level. This paper provides insights into the trade-off of precision versus analyzability at these levels.

Keywords: Energy consumption analysis · Resource usage analysis · Static analysis · Embedded systems

1 Introduction

Energy consumption and the environmental impact of computing technologies have become a major worldwide concern. It is an important issue in high-performance computing, distributed applications, and data centers. There is also

© Springer International Publishing Switzerland 2016
M. van Eekelen and U. Dal Lago (Eds.): FOPARA 2015, LNCS 9964, pp. 81–100, 2016.
DOI: 10.1007/978-3-319-46559-3_5

increased demand for complex computing systems which have to operate on batteries, such as implantable/portable medical devices or mobile phones. Despite advances in power-efficient hardware, more energy savings can be achieved by improving the way current software technologies make use of such hardware.

The process of developing energy-efficient software can benefit greatly from static analyses that estimate the energy consumed by program executions without actually running them. Such estimations can be used for different software-development tasks, such as performing automatic optimizations, verifying energy-related specifications, and helping system developers to better understand the impact of their designs on energy consumption. These tasks often relate to the source code level. On the other hand, energy consumption analysis must typically be performed at lower levels in order to take into account the effect of compiler optimizations and to link to an energy model. Thus, the inference of energy consumption information for lower levels such as the Instruction Set Architecture (ISA) or intermediate compiler representations (such as LLVM IR [19]) is fundamental for two reasons: (1) It is an intermediate step that allows propagation of energy consumption information from such lower levels up to the source code level; and (2) it enables optimizations or other applications at the ISA and LLVM IR levels.

In this paper (an improved version of [20]) we propose a static analysis approach that infers energy consumption information at the ISA and LLVM IR levels, and reflects it up to the source code level. Such information is provided in the form of *functions on input data sizes*, and is expressed by means of *assertions* that are inserted in the program representation at each of these levels. The user (i.e., the "energy-efficient software developer") can customize the system by selecting the level at which the analysis will be performed (ISA or LLVM IR) and the level at which energy information will be output (ISA, LLVM IR or source code). As we will show later, the selection of analysis level has an impact on the analysis accuracy and on the class of programs that can be analyzed.

The main goal of this paper is to study the feasibility and practicability of the proposed analysis approach and perform an initial experimental assessment to shed light on the trade-offs implied by performing the analysis at the ISA or LLVM levels. In our experiments we focus on the energy analysis of programs written in XC [31] running on the XMOS XS1-L architecture. However, the concepts presented here are neither language nor architecture dependent and thus can be applied to the analysis of other programming languages (and associated lower level program representations) and architectures as well. XC is a high-level C-based programming language that includes extensions for concurrency, communication, input/output operations, and real-time behavior. In order to potentially support different programming languages and different program representations at different levels of compilation (e.g., LLVM IR and ISA) in the same analysis framework we differentiate between the *input language* (which can be XC source, LLVM IR, or ISA) and the *intermediate semantic program representation* that the resource analysis operates on. The latter is a series of connected code blocks, represented by Horn Clauses, that we will refer to as "HC IR" from now on. We then propose a transformation from each *input language* into the HC IR and passing it to a resource analyzer. The HC IR representation as well as a transformation from LLVM IR into HC IR will be explained in Sect. 3. In our implementation we use an extension of the CiaoPP [12] resource analyzer. This analyzer always deals with the HC IR in the same way, independent of its origin, inferring energy consumption functions for all procedures in the HC IR program. The main reason for choosing Horn Clauses as the intermediate representation is that it offers a good number of features that make it very convenient for the analysis [22]. For instance, it supports naturally Static Single Assignment (SSA) and recursive forms, as will be explained later. In fact, there is a current trend favoring the use of Horn Clause programs as intermediate representations in analysis and verification tools [3,4,8,15].

Although our experiments are based on single-threaded XC programs (which do not use pointers, since XC does not support them), our claim about the generality and feasibility of our proposed approach for static resource analysis is supported by existing tools based on the Horn Clause representation that can successfully deal with C source programs that exhibit interesting features such as the use of pointers, arrays, shared-memory, or concurrency in order to analyze and verify a wide range of properties [8,10,15]. For example [10] is a tool for the verification of safety properties of C programs which can reason about scalars and pointer addresses, as well as memory contents. It represents the bytecode corresponding to a C program by using (constraint) Horn clauses.

Both static analysis and energy models can potentially relate to any language level (such as XC source, LLVM IR, or ISA). Performing the analysis at a given level means that the representation of the program at that level is transformed into the HC IR, and the analyzer "mimics" the semantics of instructions at that level. The energy model at a given level provides basic information on the energy cost of instructions at that level. The analysis results at a given level can be mapped upwards to a higher level, e.g. from ISA or LLVM IR to XC. Furthermore, it is possible to perform analysis at a given level with an energy model for a lower level. In this case the energy model must be reflected up to the analysis level.

Our hypothesis is that the choice of level will affect the accuracy of the energy models and the precision of the analysis in opposite ways: energy models at lower levels (e.g. at the ISA level) will be more precise than at higher levels (e.g. XC source code), since the closer to the hardware, the easier it is to determine the effect of the execution on the hardware. However, at lower levels more program structure and data type/shape information is lost due to lower-level representations, and we expect a corresponding loss of analysis accuracy (without using complex techniques for recovering type information and abstracting memory operations). This hypothesis about the analysis/modelling level trade-off (and potential choices) is illustrated in Fig. 1. The possible choices are classified into two groups: those that analyze and model at the same level, and those that operate at different levels. For the latter, the problem is finding good mappings between software segments from the level at which the model is defined up to

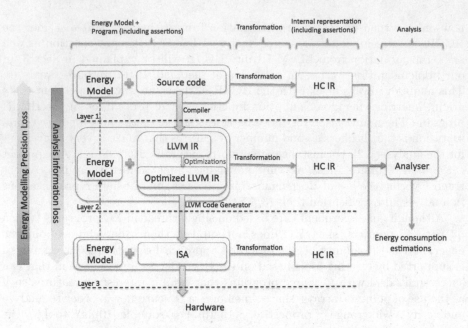

Fig. 1. Analysis/modelling level trade-off and potential choices.

the level at which the analysis is performed, in a way that does not lose accuracy in the energy information.

In this paper we concentrate on two of these choices and their comparison, to see if our hypothesis holds. In particular, the first approach (choice 1) is represented by analysing the generated ISA-level code using models defined at the ISA level that express the energy consumed by the execution of individual ISA instructions. This approach was explored in [21]. It used the precise ISA-level energy models presented in [17], which when used in the static analysis of [21] for a number of small numerical programs resulted in the inference of functions that provide reasonably accurate energy consumption estimations for any input data size (3.9 % average error vs. hardware measurements). However, when dealing with programs involving structured types such as arrays, it also pointed out that, due to the loss of information related to program structure and types of arguments at the ISA level (since it is compiled away and no longer relates cleanly to source code), the power of the analysis was limited. In this paper we start by exploring an alternative approach: the analysis of the generated LLVM IR (which retains much more of such information, enabling more direct analysis as well as mapping of the analysis information back to source level) together with techniques that map segments of ISA instructions to LLVM IR blocks [7] (choice 2). This mapping is used to propagate the energy model information defined at the ISA level up to the level at which the analysis is performed, the LLVM IR level. In order to complete the LLVM IR-level analysis, we have also developed and implemented a transformation from LLVM IR into HC IR and used the CiaoPP resource analyzer. This results in a parametric analysis that similarly to [21] infers energy consumption functions, but operating on the LLVM IR level rather than the ISA level.

We have performed an experimental comparison of the two choices for generating energy consumption functions. Our results support our intuitions about the trade-offs involved. They also provide evidence that the LLVM IR-level analysis (choice 2) offers a good compromise within the level hierarchy, since it broadens the class of programs that can be analyzed without significant loss of accuracy.

In summary, the original contributions of this paper are:

1. A translation from LLVM IR to HC IR (Sect. 3).
2. The integration of all components into an experimental tool architecture, enabling the static inference of energy consumption information in the form of *functions on input data sizes* and the experimentation with the trade-offs described above (Sect. 2). The components are: LLVM IR and ISA translations, ISA-level energy model and mapping technique (Sect. 4 and [7,17]), and analysis tools (Sect. 5 and [25,28]).
3. The experimental results and evidence of trade-off of precision versus analyzability (Sect. 6).
4. A sketch of how the static analysis system can be integrated in a source-level Integrated Development Environment (IDE) (Sect. 2).

Finally, some related work is discussed in Sects. 7 and 8 summarises our conclusions and comments on ongoing and future work.

2   Overview of the Analysis at the LLVM IR Level

An overview of the proposed analysis system at the LLVM IR level using models at the ISA level is depicted in Fig. 2. The system takes as input an XC source program that can (optionally) contain assertions (used to provide useful hints and information to the analyzer), from which a *Transformation and Mapping* process (dotted red box) generates first its associated LLVM IR using the xcc compiler. Then, a transformation from LLVM IR into HC IR is performed (explained in Sect. 3) obtaining the intermediate representation (green box) that is supplied to the CiaoPP analyzer. This representation includes assertions that express the energy consumed by the LLVM IR blocks, generated from the information produced by the mapper tool (as explained in Sect. 4). The *CiaoPP analyzer* (blue box, described in Sect. 5) takes the HC IR, together with the assertions which express the energy consumed by LLVM IR blocks, and possibly some additional (trusted) information, and processes them, producing the analysis results, which are expressed also using assertions. Based on the procedural interpretation of these HC IR programs and the resource-related information contained in the assertions, the resource analysis can infer static bounds on the energy consumption of the HC IR programs that are applicable to the original LLVM IR and, hence, to their corresponding XC programs. The analysis results include energy consumption information expressed as functions on data sizes for the whole

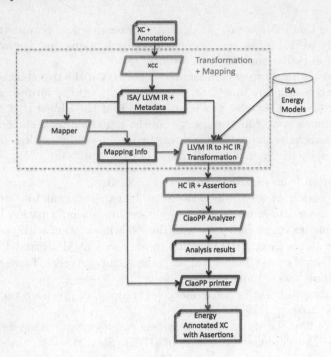

Fig. 2. An overview of the analysis at the LLVM IR level using ISA models. (Color figure online)

program and for all the procedures and functions in it. Such results are then processed by the *CiaoPP printer* (purple box) which presents the information to the program developer in a user-friendly format.

3   LLVM IR to HC IR Transformation

In this section we describe the LLVM IR to HC IR transformation that we have developed in order to achieve the complete analysis system at the LLVM IR level proposed in the paper (as already mentioned in the overview given in Sect. 2 and depicted in Fig. 2).

A Horn clause (HC) is a first-order predicate logic formula of the form $\forall (S_1 \wedge \ldots \wedge S_n \rightarrow S_0)$ where all variables in the clause are universally quantified over the whole formula, and $S_0, S_1, \ldots, S_n$ are atomic formulas, also called literals. It is usually written $S_0 :- S_1, \ldots, S_n$.

The HC IR representation consists of a sequence of *blocks* where each block is represented as a *Horn clause*:

$$< block\_id > (< params >) :- \ S_1, \ \ldots, S_n.$$

Each block has an entry point, that we call the *head* of the block (to the left of the :− symbol), with a number of parameters $< params >$, and a sequence of steps (the *body*, to the right of the : − symbol). Each of these S_i steps (or *literals*) is either (the representation of) an LLVM IR *instruction*, or a *call* to another (or the same) block. The transformation ensures that the program information relevant to resource usage is preserved, so that the energy consumption functions of the HC IR programs inferred by the resource analysis are applicable to the original LLVM IR programs.

The transformation also passes energy values for the LLVM IR level for different programs based on the ISA/LLVM IR mapping information that express the energy consumed by the LLVM IR blocks, as explained in Sect. 4. Such information is represented by means of *trust* assertions (in the Ciao assertion language [13]) that are included in the HC IR. In general, *trust* assertions can be used to provide information about the program and its constituent parts (e.g., individual instructions or whole procedures or functions) to be trusted by the analysis system, i.e., they provide base information assumed to be true by the inference mechanism of the analysis in order to propagate it throughout the program and obtain information for the rest of its constituent parts.

LLVM IR programs are expressed using typed assembly-like instructions. Each function is in SSA form, represented as a sequence of basic blocks. Each basic block is a sequence of LLVM IR instructions that are guaranteed to be executed in the same order. Each block ends in either a branching or a return instruction. In order to transform an LLVM IR program into the HC IR, we follow a similar approach as in a previous ISA-level transformation [21]. However, the LLVM IR includes an additional type transformation as well as better memory modelling.

The following subsections describe the main aspects of the transformation.

3.1 Inferring Block Arguments

As described before, a *block* in the HC IR has an entry point (head) with input/output parameters, and a body containing a sequence of steps (here, representations of LLVM IR instructions). Since the scope of the variables in LLVM IR blocks is at the function level, the blocks are not required to pass parameters while making jumps to other blocks. Thus, in order to represent LLVM IR blocks as HC IR blocks, we need to infer input/output parameters for each block.

For entry blocks, the input and output arguments are the same as the ones to the function. We define the functions $param_{in}$ and $param_{out}$ which infer input and output parameters to a block respectively. These are recomputed according to the following definitions until a fixpoint is reached:

$$params_{out}(b) = (kill(b) \ \cup \ params_{in}(b)) \ \cap \ \bigcup_{b' \in next(b)} params_{out}(b')$$

$$params_{in}(b) = gen(b) \ \cup \ \bigcup_{b' \in next(b)} params_{in}(b')$$

where $next(b)$ denotes the set of immediate target blocks that can be reached from block b with a jump instruction, while $gen(b)$ and $kill(b)$ are the read and written variables in block b respectively, which are defined as:

$$kill(b) = \bigcup_{k=1}^{n} def(k)$$

$$gen(b) = \bigcup_{k=1}^{n} \{v \mid v \in ref(k) \wedge \forall(j < k).v \notin def(j)\}$$

where $def(k)$ and $ref(k)$ denote the variables written or referred to at a node (instruction) k in the block, respectively, and n is the number of nodes in the block.

Note that the LLVM IR is in SSA form at the function level, which means that blocks may have ϕ nodes which are created while transforming the program into SSA form. A ϕ node is essentially a function defining a new variable by selecting one of the multiple instances of the same variable coming from multiple predecessor blocks:

$$x = \phi(x_1, x_2, ..., x_n)$$

def and ref for this instruction are $\{x\}$ and $\{x_1, x_2, ..., x_n\}$ respectively. An interesting feature of our approach is that ϕ nodes are not needed. Once the input/output parameters are inferred for each block as explained above, a post-process gets rid of all ϕ nodes by modifying block input arguments in such a way that blocks receive x directly as an input and an appropriate x_i is passed by the call site. This will be illustrated later in Sect. 3.3.

Consider the example in Fig. 4 (left), where the LLVM IR block *looptest* is defined. The body of the block reads from 2 variables without previously defining them in the same block. The fixpoint analysis would yield:

$$params_{in}(looptest) = \{Arr, I\}$$

which is used to construct the HC IR representation of the *looptest* block shown in Fig. 4 (right), line 3.

3.2   Translating LLVM IR Types into HC IR Types

LLVM IR is a typed representation which allows retaining much more of the (source) program information than the ISA representation (e.g., types defining compound data structures). Thus, we define a mechanism to translate LLVM IR types into their counterparts in HC IR.

The LLVM type system defines primitive and derived types. The primitive types are the fundamental building blocks of the type system. Primitive types include *label, void, integer, character, floating point, x86mmx,* and *metadata.* The *x86mmx* type represents a value held in an MMX register on an x86 machine and the *metadata* type represents embedded metadata. The derived types are created from primitive types or other derived types. They include *array, function, pointer, structure, vector, opaque.* Since the XCore platform supports neither pointers nor floating point data types, the LLVM IR code generated from XC programs uses only a subset of the LLVM types.

At the HC IR level we use *regular types*, one of the type systems supported by CiaoPP [12]. Translating LLVM IR primitive types into regular types is straightforward. The *integer* and *character* types are abstracted as *num* regular type, whereas the *label*, *void*, and *metadata* types are represented as *atm* (atoms).

For derived types, corresponding non-primitive regular types are constructed during the transformation phase. Supporting non-primitive types is important because it enables the analysis to infer energy consumption functions that depend on the sizes of internal parts of complex data structures. The array, vector, and structure types are represented as follows:

$$array\_type \rightarrow (nested)list$$
$$vector\_type \rightarrow (nested)list$$
$$structure\_type \rightarrow functor\_term$$

Both the *array* and *vector* types are represented by the *list* type in CiaoPP which is a special case of compound term. The type of the elements of such lists can be again a primitive or a derived type. The *structure* type is represented by a compound term which is composed of an atom (called the *functor*, which gives a name to the structure) and a number of *arguments*, which are again either primitive or derived types. LLVM also introduces pointer types in the intermediate representation, even if the front-end language does not support them (as in the case of XC, as mentioned before). Pointers are used in the pass-by-reference mechanism for arguments, in memory allocations in *alloca* blocks, and in memory load and store operations. The types of these pointer variables in the HC IR are the same as the types of the data these pointers point to.

```
struct mystruct{                              :- regtype array1/1.
  int x;                                      array1:=[] | [~struct|array1].
  int arr[5];
};                                            :- regtype struct/1.

void print(struct mystruct [] Arg, int N)     struct:=mystruct(~num,~array2).
{                                             
  ...                                         :- regtype array2/1.
}                                             array2:=[] | [~num|array2].
```

Fig. 3. An XC program and its type transformation into HC IR.

Consider for example the types in the XC program shown in Fig. 3. The type of argument *Arg* of the *print* function is an array of *mystruct* elements. *mystruct* is further composed of an integer and an array of integers. The LLVM IR code generated by xcc for the function signature *print* in Fig. 3 (left) is:

define void @print($[0 \times \{i32, [5 \times i32]\}] * $ noalias nocapture)

The function argument type in the LLVM IR ($[0 \times \{i32, [5 \times i32]\}]$) is the typed representation of the argument *Arg* to the function in the XC program.

It represents an array of arbitrary length with elements of $\{i32, [5 \times i32]\}$ structure type which is further composed of an $i32$ integer type and a $[5 \times i32]$ array type, i.e., an array of 5 elements of $i32$ integer type.[1]

This type is represented in the HC IR using the set of regular types illustrated in Fig. 3 (right). The regular type $array1$, is a list of *struct* elements (which can also be simply written as `array1 := list(struct)`). Each *struct* type element is represented as a functor $mystruct/2$ where the first argument is a *num* and the second is another list type $array2$. The type $array2$ is defined to be a list of *num* (which, again, can also be simply written as `array2 := list(num)`).

3.3   Transforming LLVM IR Blocks/Instructions into HC IR

In order to represent an LLVM IR function by an HC IR function (i.e., a predicate), we need to represent each LLVM IR block by an HC IR block (i.e., a Horn clause) and hence each LLVM IR instruction by an HC IR literal.

```
1   alloca:                              1   alloca(N, Arr):−
2    br label looptest                   2    looptest(N, Arr).
3   looptest:                            3   looptest(I, Arr):−
4    %I=phi i32[%N,%alloca],             4    icmp_ne(I, 0, Zcmp),
                  [%I1,%loopbody]         5    loopbody_loopend(Zcmp,I,Arr).
5    %Zcmp=icmp ne i32 %I, 0             6   icmp_ne(X, Y, 1):− X \= Y.
6    br i1 %Zcmp, label %loopbody,       7   icmp_ne(X, Y, 0):− X = Y.
                  label %loopend         8   loopbody_loopend(Zcmp,I,Arr):−
7   loopbody:                            9    Zcmp=1,
8    %Elm=getelementptr [0xi32]*%Arr,   10    nth(I, Arr, Elm),
                  i32 0,i32 %I          11    //process list element 'Elm'
9    //process array element 'Elm'      12    I1 is I − 1, sub(I,1,I1),
10   %I1=sub i32 %I, 1                  13    looptest(I1, Arr).
11   br label %looptest                 14   loopbody_loopend(Zcmp,I,Arr):−
12  loopend:                            15    Zcmp=0.
13   ret void
```

Fig. 4. LLVM IR Array traversal example (left) and its HC IR representation (right)

The LLVM IR instructions are transformed into equivalent HC IR literals where the semantics of the execution of the LLVM IR instructions are either described using trust assertions or by giving definition to HC IR literals. The *phi assignment* instructions are removed and the semantics of the *phi assignment* are preserved on the call sites. For example, the *phi assignment* is removed from the HC IR block in Fig. 4 (right) and the semantics of the *phi assignment* is preserved on the call sites of the *looptest* (lines 2 and 14). The call sites *alloca* (line 2) and *loopbody* (line 13) pass the corresponding value as an argument to *looptest*, which is received by *looptest* in its first argument I.

Consider the instruction *getelementptr* at line 8 in Fig. 4 (left), which computes the address of an element of an array $\%Arr$ indexed by $\%I$ and assigns it to a variable $\%Elm$. Such an instruction is represented by a call to an abstract predicate $nth/3$, which extracts a reference to an element from a list, and whose

---

[1] $[0 \times i32]$ specifies an arbitrary length array of $i32$ integer type elements.

effect of execution on energy consumption as well as the relationship between the sizes of input and output arguments is described using trust assertions. For example, the assertion:

```
:- trust pred nth(I, L, Elem)
   :(num(I), list(L, num), var(Elem))
  => ( num(I), list(L, num), num(Elem),
       rsize(I, num(IL, IU)),
       rsize(L, list(LL, LU, num(EL, EU))),
       rsize(Elem, num(EL, EU)) )
     + (resource(avg, energy, 1215439) ).
```

indicates that if the nth(I, L, Elem) predicate (representing the *getelementptr* LLVM IR instruction) is called with I and L bound to an integer and a list of numbers respectively, and *Elem* an unbound variable (precondition field ":"), then, after the successful completion of the call (postcondition field "=>"), *Elem* is an integer number and the lower and upper bounds on its size are equal to the lower and upper bounds on the sizes of the elements of the list L. The sizes of the arguments to *nth/3* are expressed using the property *rsize* in the assertion language. The lower and upper bounds on the length of the list L are LL and LU respectively. Similarly, the lower and upper bounds on the elements of the list are EL and EU respectively, which are also the bounds for *Elem*. The *resource* property (global computational properties field +) expresses that the energy consumption for the instruction is an average value (1215439 nano-joules[2]).

The branching instructions in LLVM IR are transformed into calls to target blocks in HC IR. For example, the branching instruction at line 6 in Fig. 4 (left), which jumps to one of the two blocks *loopbody* or *loopend* based on the Boolean variable *Zcmp*, is transformed into a call to a predicate with two clauses (line 5 in Fig. 4 (right)). The name of the predicate is the concatenation of the names of the two LLVM IR blocks mentioned above. The two clauses of the predicate defined at lines 8–13 and 14–15 in Fig. 4 (right) represent the LLVM IR blocks *loopbody* and *loopend* respectively. The test on the conditional variable is placed in both clauses to preserve the semantics of the conditional branch.

4 Obtaining the Energy Consumption of LLVM IR Blocks

Our approach requires producing assertions that express the energy consumed by each call to an LLVM IR block (or parts of it) when it is executed. To achieve this we take as starting point the energy consumption information available from an existing XS1-L ISA Energy Model produced in our previous work of ISA level analysis [21] using the techniques described in [17]. We refer the reader to [17] for a detailed study of the energy consumption behaviour of the XS1-L architecture, containing a description of the test and measurement process along with the construction and full evaluation of such model. In the experiments performed in this paper a single, constant energy value is assigned to each instruction in the ISA based on this model.

---

[2] nJ, 10^{-9} joules.

A mechanism is then needed to propagate such ISA-level energy information up to the LLVM IR level and obtain energy values for LLVM IR blocks. A set of mapping techniques serve this purpose by creating a fine-grained mapping between segments of ISA instructions and LLVM IR code segments, in order to enable the energy characterization of each LLVM IR instruction in a program, by aggregating the energy consumption of the ISA instructions mapped to it. Then, the energy value assigned to each LLVM IR block is obtained by aggregating the energy consumption of all its LLVM IR instructions. The mapping is done by using the debug mechanism where the debug information, preserved during the lowering phase of the compilation from LLVM IR to ISA, is used to track ISA instructions against LLVM IR instructions. A full description and formalization of the mapping techniques is given in [7].

5   Resource Analysis with CiaoPP

In order to perform the global energy consumption analysis, our approach leverages the CiaoPP tool [12], the preprocessor of the Ciao programming environment [13]. CiaoPP includes a global static analyzer which is parametric with respect to resources and type of approximation (lower and upper bounds) [25,28]. The framework can be instantiated to infer bounds on a very general notion of resources, which we adapt in our case to the inference of energy consumption. In CiaoPP, a resource is a user-defined *counter* representing a (numerical) non-functional global property, such as execution time, execution steps, number of bits sent or received by an application over a socket, number of calls to a predicate, number of accesses to a database, etc. The instantiation of the framework for energy consumption (or any other resource) is done by means of an assertion language that allows the user to define resources and other parameters of the analysis by means of assertions. Such assertions are used to assign basic resource usage functions to elementary operations and certain program constructs of the base language, thus expressing how the execution of such operations and constructs affects the usage of a particular resource. The resource consumption provided can be a constant or a function of some input data values or sizes. The same mechanism is used as well to provide resource consumption information for procedures from libraries or external code when code is not available or to increase the precision of the analysis.

For example, in order to instantiate the CiaoPP general analysis framework for estimating bounds on energy consumption, we start by defining the identifier ("counter") associated to the energy consumption resource, through the following Ciao declaration:

```
:- resource energy.
```

We then provide assertions for each HC IR block expressing the energy consumed by the corresponding LLVM IR block, determined from the energy model, as explained in Sect. 4. Based on this information, the global static analysis can then infer bounds on the resource usage of the whole program (as well as procedures and functions in it) as functions of input data sizes. A full description of how this is done can be found in [28].

Consider the example in Fig. 4 (right). Let P_e denote the energy consumption function for a predicate P in the HC IR representation (set of blocks with the same name). Let c_b represent the energy cost of an LLVM IR block b. Then, the inferred equations for the HC IR blocks in Fig. 4 (right) are:

$$alloca_e(N, Arr) = c_{alloca} + looptest_e(N, Arr)$$

$$looptest_e(N, Arr) = c_{looptest} + loopbody\_loopend_e(0 \neq N, N, Arr)$$

$$loopbody\_loopend_e(B, N, Arr) = \begin{cases} looptest_e(N - 1, Arr) & \text{if } B \text{ is true} \\ + c_{loopbody} & \\ c_{loopend} & \text{if } B \text{ is false} \end{cases}$$

If we assume (for simplicity of exposition) that each LLVM IR block has unitary cost, i.e., $c_b = 1$ for all LLVM IR blocks b, solving the above recurrence equations, we obtain the energy consumed by `alloca` as a function of its input data size (N):

$$alloca_e(N, Arr) = 2 \times N + 3$$

Note that using average energy values in the model implies that the energy function for the whole program inferred by the upper-bound resource analysis is an approximation of the actual upper bound (possibly below it). Thus, theoretically, to ensure that the analysis infers an upper bound, we need to use upper bounds as well in the energy models. This is not a trivial task as the worst case energy consumption depends on the data processed, is likely to be different for different instructions, and unlikely to occur frequently in subsequent instructions. A first investigation into the effect of different data on the energy consumption of individual instructions, instruction sequences and full programs is presented in [26]. A refinement of the energy model to capture upper bounds for individual instructions, or a selected subset of instructions, is currently being investigated, extending the first experiments into the impact of data into worst case energy consumption at instruction level as described in Sect. 5.5 of [17].

6 Experimental Evaluation

We have performed an experimental evaluation of our techniques on a number of selected benchmarks. Power measurement data was collected for the XCore platform by using appropriately instrumented power supplies, a power-sense chip, and an embedded system for controlling the measurements and collecting the power data. Details about the power monitoring setup used to run our benchmarks and measure their energy consumption can be found in [17]. The main goal of our experiments was to shed light on the trade-offs implied by performing the analysis at the ISA level (without using complex mechanisms for propagating type information and representing memory) and at the LLVM level using models defined at the ISA level together with a mapping mechanism.

There are two groups of benchmarks that we have used in our experimental study. The first group is composed of four small recursive numerical programs that have a variety of user defined functions, arguments, and calling patterns (first four benchmarks in Table 2). These benchmarks only operate over primitive data types and do not involve any structured types. The second group of benchmarks (the last five benchmarks in Table 2) differs from the first group in the sense that they all involve structured types. These are recursive or iterative.

The second group of benchmarks includes fir(N) and biquad(N). The former is a (finite impulse response) filter program, which attenuates or amplifies one specific frequency range of a given input signal. It computes the inner-product of two vectors: a vector of input samples, and a vector of coefficients. The latter is an equaliser, which takes a signal and attenuates/amplifies different frequency bands. It uses a cascade of Biquad filters where each filter attenuates or amplifies one specific frequency range. The energy consumed depends on the number of banks N.

None of the XC benchmarks contain any assertions that provide information to help the analyzer. Table 1 shows detailed experimental results. Column **SA energy function** shows the energy consumption functions, which depend on input data sizes, inferred for each program by the static analyses performed at the ISA and LLVM IR levels (denoted with subscripts *isa* and *llvm* respectively). We can see that the analysis is able to infer different kinds of functions (polynomial, exponential, etc.). Column **HW** shows the actual energy consumption in nano-joules measured on the hardware corresponding to the execution of the programs with input data of different sizes (shown in column **Input Data Size**). **Estimated** presents the energy consumption estimated by static analysis. This is obtained by evaluating the functions in column **SA energy function** for the input data sizes in column **Input Data Size**. The value N/A in such column means that the analysis has not been able to infer any useful energy consumption function and, thus, no estimated value is obtained. Column **Err vs. HW** shows the error of the values estimated by the static analysis with respect to the actual energy consumption measured on the hardware, calculated as follows: **Err vs. HW** $= (\frac{\text{LLVM}(or\ \text{ISA})-\text{HW}}{\text{HW}} \times 100)\%$. Finally, the last column shows the ratio between the estimations of the analysis at the ISA and LLVM IR levels.

Table 2 shows a summary of results. The first two columns show the name and short description of the benchmarks. The columns under **Err vs. HW** show the average error obtained from the values given in Table 1 for different input data sizes. The last row of the table shows the average error over the number of benchmarks analyzed at each level.

The experimental results show that:

- For the benchmarks in the first group, both the ISA- and LLVM IR-level analyses are able to infer useful energy consumption functions. On average, the analysis performed at either level is reasonably accurate and the relative error between the two analyses at different levels is small. ISA-level estimations are slightly more accurate than the ones at the LLVM IR level (3.9 % vs. 9 % error on average with respect to the actual energy consumption measured on

Table 1. Comparison of the accuracy of energy analyses at the LLVM IR and ISA levels.

SA energy function (nJ)	Input Size	HW (nJ)	Estimated (nJ) llvm	isa	Err vs. HW% llvm	isa	isa/ llvm
$Fact_{isa}(N)=$	N=8	227	237	212	4.6	-6.4	0.9
24.26 N + 18.43	N=16	426	453	406	6.5	-4.5	0.9
$Fact_{llvm}(N)=$	N=32	824	886	794	7.6	-3.5	0.9
27.03 N + 21.28	N=64	1690	1751	1571	3.6	-7.0	0.9
$Fib_{isa}(N)^a$=26.88$fib(N)$	N=2	75	74	65	-1.16	-12.5	0.89
+22.85 $lucas(N)^b$−30.04	N=4	219	241	210	10	-4.1	0.87
	N=8	1615	1853	1608	14.75	-0.4	0.87
$Fib_{llvm}(N)^a$ =32.5$fib(N)$	N=15	47 × 10³	54 × 10³	47 × 10³	16.47	1.2	0.87
+25.6 $lucas(N)^b$ − 35.65	N=26	9.30 × 10⁶	10.9 × 10⁶	9.5 × 10⁶	17.3	1.74	0.87
$Sqr_{isa}(N)=$	N=9	1242	1302	1148	4.8	-7.5	0.88
8.6 N^2 + 48.7 N + 15.6	N=27	8135	8734	7579	7.4	-6.8	0.87
	N=73	52 × 10³	57 × 10³	49 × 10³	8.5	-6.5	0.86
$Sqr_{llvm}(N)=$	N=144	19.7 × 10⁴	21.4 × 10⁴	18.4 × 10⁴	8.89	-6.4	0.86
10 N^2 + 53 N + 15.6	N=234	51 × 10⁴	56 × 10⁴	48 × 10⁴	9.61	-5.86	0.86
	N=360	11.89 × 10⁵	13 × 10⁵	11.2 × 10⁵	10.49	-5.16	0.86
	N=3	326	344	3.6	5.7	-6.0	0.89
$PowerOfTwo_{isa}(N)=$	N=6	2729	2965	2631	8.7	3.6	0.89
41.5 × 2^N − 25.9	N=9	21.9 × 10³	23.9 × 10³	21.2 × 10³	9	3.3	0.89
$PowerOfTwo_{llvm}(N) =$	N=12	17.57 × 10⁴	19.1 × 10⁴	17 × 10⁴	9	-3.3	0.89
46.8 × 2^N − 29.9	N=15	13.8 × 10⁵	15.3 × 10⁵	13.6 × 10⁵	11	-1.5	0.89
	N=57	1138	1179	N/A	3.60	N/A	N/A
$reverse_{llvm}(N)=$	N=160	3125	3185	N/A	1.91	N/A	N/A
19.47 N + 69.33	N=320	6189	6301	N/A	1.82	N/A	N/A
	N=720	13848	14092	N/A	1.76	N/A	N/A
	N=1280	24634	24998	N/A	1.48	N/A	N/A
	N=5	7453	7569	N/A	-2	N/A	N/A
$matmult_{llvm}(N)=$	N=15	15.79 × 10⁴	15.9 × 10⁴	N/A	1.03	N/A	N/A
42.47 N^3 + 68.85 N^2+	N=20	36.29 × 10⁴	36.8 × 10⁴	N/A	1.51	N/A	N/A
49.9 N + 24.22	N=25	69.56 × 10⁴	70.8 × 10⁴	N/A	1.77	N/A	N/A
	N=31	13.07 × 10⁵	13.3 × 10⁵	N/A	1.98	N/A	N/A
	N=131; M=69	14.5 × 10³	13.2 × 10³	N/A	8.65	N/A	N/A
$concat_{llvm}(N, M)=$	N=170; M=182	25.44 × 10³	23.3 × 10³	N/A	8.60	N/A	N/A
65.7 N + 65.7 M + 137	N=188; M=2	13.8 × 10³	12.6 × 10³	N/A	8.59	N/A	N/A
	N=13; M=134	10.7 × 10³	9.79 × 10³	N/A	8.74	N/A	N/A
$biquad_{llvm}(N)=$	N=5	871	836	N/A	-4	N/A	N/A
157 N + 51.7	N=7	1187	1151	N/A	-3.1	N/A	N/A
	N=10	1660	1622	N/A	-2.31	N/A	N/A
	N=14	2290	2250	N/A	-1.75	N/A	N/A
$fir_{llvm}(N)=$	N=85	2999	2839	N/A	-5.3	N/A	N/A
31.8 N + 137	N=97	3404	3221	N/A	-5.37	N/A	N/A
	N=109	3812	3602	N/A	-5.5	N/A	N/A
	N=121	4227	3984	N/A	-5.7	N/A	N/A

[a] It uses mathematical functions fib and $lucas$, a function expansion would yield:
$Fib_{isa}(N)$=34.87 × 1.62^N + 10.8 × (−0.62)^N − 30
$Fib_{llvm}(N)$=40.13 × 1.62^N + 11.1 × (−0.62)^N − 35.65
[b] $Lucas(n)$ satisfy the recurrence relation $L_n = L_{n-1} + L_{n-2}$ with $L_1 = 1, L_2 = 3$

Table 2. LLVM IR- vs. ISA-level analysis accuracy.

Program	Description	Err vs. HW		isa/llvm
		llvm	isa	
fact(N)	Calculates N!	5.6 %	5.3 %	0.89
fibonacci(N)	Nth Fibonacci number	11.9 %	4 %	0.87
sqr(N)	Computes N^2 performing additions	9.3 %	3.1 %	0.86
pow_of_two(N)	Calculates 2^N without multiplication	9.4 %	3.3 %	0.89
Average		**9 %**	**3.9 %**	**0.92**
reverse(N, M)	Reverses an array	2.18 %	N/A	N/A
concat(N, M)	Concatenation of arrays	8.71 %	N/A	N/A
matmult(N, M)	Matrix multiplication	1.47 %	N/A	N/A
fir(N)	Finite Impulse Response filter	5.47 %	N/A	N/A
biquad(N)	Biquad equaliser	3.70 %	N/A	N/A
Average		**3.0 %**	**N/A**	**N/A**
Overall average		**6.4 %**	**3.9 %**	**0.92**

the hardware, respectively). This is because the ISA-level analysis uses very accurate energy models, obtained from measuring directly at the ISA level, whereas at the LLVM IR level, such ISA-level model needs to be propagated up to the LLVM IR level using (approximated) mapping information. This causes a slight loss of accuracy.

– For the second group of benchmarks, the ISA level analysis is not able to infer useful energy functions. This is due to the fact that significant program structure and data type/shape information is lost due to lower-level representations, which sometimes makes the analysis at the ISA level very difficult or impossible. In order to overcome this limitation and improve analysis accuracy, significantly more complex techniques for recovering type information and representing memory in the HC IR would be needed. In contrast, type-/shape information is preserved at the LLVM IR level, which allows analyzing programs using data structures (e.g., arrays). In particular, all the benchmarks in the second group are analyzed at the LLVM IR level with reasonable accuracy (3 % error on average). In this sense, the LLVM IR-level analysis is more powerful than the one at the ISA level. The analysis is also reasonably efficient, with analysis times of about 5 to 6 seconds on average, despite the naive implementation of the interface with external recurrence equation solvers, which can be improved significantly. The scalability of the analysis follows from the fact that it is compositional and can be performed in a modular way, making use of the Ciao assertion language to store results of previously analyzed modules.

7 Related Work

Few papers can be found in the literature focusing on static analysis of energy consumption. A similar approach to the one presented in this paper and our previously developed analysis [21] (from which it builds on) was proposed for upper-bound energy analysis of Java bytecode programs in [24], where the Jimple (a typed three-address code) representation of Java bytecode was transformed into Horn Clauses, and a simple energy model at the Java bytecode level [18] was used. However, this work did not compare the results with actual, measured energy consumption.

In all the approaches mentioned above, instantiations for energy consumption of general resource analyzers are used, namely [25] in [24] and [21], and [28] in this paper. Such resource analyzers are based on setting up and solving recurrence equations, an approach proposed by Wegbreit [32] that has been developed significantly in subsequent work [1,5,6,25,27,28,30]. Other approaches to static analysis based on the transformation of the analyzed code into another (intermediate) representation have been proposed for analyzing low-level languages [11] and Java (by means of a transformation into Java bytecode) [2]. In [2], cost relations are inferred directly for these bytecode programs, whereas in [24] the bytecode is first transformed into Horn Clauses. The general resource analyzer in [25] was also instantiated in [23] for the estimation of execution times of logic programs running on a bytecode-based abstract machine. The approach used timing models at the bytecode instruction level, for each particular platform, and program-specific mappings to lift such models up to the Horn Clause level, at which the analysis was performed.

By contrast to the generic approach based on CiaoPP, an approach operating directly on the LLVM IR representation is explored in [9]. Though relying on similar analysis techniques, the approach can be integrated more directly in the LLVM toolchain and is in principle applicable to any languages targeting this toolchain. The approach uses the same LLVM IR energy model and mapping technique as the one applied in this paper.

A number of static analyses are also aimed at worst case execution time (WCET), usually for imperative languages in different application domains (see e.g., [33] and its references). The worst-case analysis presented in [16], which is not based on recurrence equation solving, distinguishes instruction-specific (not proportional to time, but to data) from pipeline-specific (roughly proportional to time) energy consumption. However, in contrast to the work presented here and in [23], these worst case analysis methods do not infer cost functions on input data sizes but rather absolute maximum values, and they generally require the manual annotation of loops to express an upper-bound on the number of iterations. An alternative approach to WCET was presented in [14]. It is based on the idea of amortisation, which allows to infer more accurate yet safe upper bounds by averaging the worst execution time of operations over time. It was applied to a functional language, but the approach is in principle generally applicable. A timing analysis based on game-theoretic learning was presented in [29]. The approach combines static analysis to find a set of basic paths which are then tested. In principle, such approach could be adapted to infer energy usage. Its main advantage is that this analysis can infer distributions on time, not only average values.

8   Conclusions and Future Work

We have presented techniques for extending to the LLVM IR level our tool chain for estimating energy consumption as functions on program input data sizes. The approach uses a mapping technique that leverages the existing debugging mechanisms in the XMOS XCore compiler tool chain to propagate an ISA-level energy model to the LLVM IR level. A new transformation constructs a block representation that is supplied, together with the propagated energy values, to a parametric resource analyzer that infers the program energy cost as functions on the input data sizes.

Our results suggest that performing the static analysis at the LLVM IR level is a reasonable compromise, since (1) LLVM IR is close enough to the source code level to preserve most of the program information needed by the static analysis, and (2) the LLVM IR is close enough to the ISA level to allow the propagation of the ISA energy model up to the LLVM IR level without significant loss of accuracy for the examples studied. Our experiments are based on single-threaded programs. We also have focused on the study of the energy consumption due to computation, so that we have not tested programs where storage and networking is important. However, this could potentially be done in future work, by using the CiaoPP static analysis, which already infers bounds on data sizes, and combining such information with appropriate energy models of communication and storage. It remains to be seen whether the results would carry over to other classes of programs, such as multi-threaded programs and programs where timing is more important. In this sense our results are preliminary, yet they are promising enough to continue research into analysis at LLVM IR level and into ISA-LLVM IR energy mapping techniques to enable the analysis of a wider class of programs, especially multi-threaded programs.

Acknowledgements. This research has received funding from the European Union 7th Framework Program agreement no 318337, ENTRA, Spanish MINECO TIN'12-39391 *StrongSoft* project, and the Madrid M141047003 *N-GREENS* program.

References

1. Albert, E., Arenas, P., Genaim, S., Puebla, G.: Closed-form upper bounds in static cost analysis. J. Autom. Reasoning **46**(2), 161–203 (2011)
2. Albert, E., Arenas, P., Genaim, S., Puebla, G., Zanardini, D.: Cost analysis of Java bytecode. In: Nicola, R. (ed.) ESOP 2007. LNCS, vol. 4421, pp. 157–172. Springer, Heidelberg (2007). doi:10.1007/978-3-540-71316-6_12
3. Bjørner, N., Fioravanti, F., Rybalchenko, A., Senni, V. (eds.) Proceedings of First Workshop on Horn Clauses for Verification and Synthesis, vol. 169. EPTCS, July 2014

4. Moura, L., Bjørner, N.: Z3: an efficient SMT solver. In: Ramakrishnan, C.R., Rehof, J. (eds.) TACAS 2008. LNCS, vol. 4963, pp. 337–340. Springer, Heidelberg (2008). doi:10.1007/978-3-540-78800-3_24
5. Debray, S.K., Lin, N.-W., Hermenegildo, M.: Task granularity analysis in logic programs. In: Proceeding of the 1990 ACM Conference on Programming Language Design and Implementation, pp. 174–188. ACM Press, June 1990
6. Debray, S.K., López-García, P., Hermenegildo, M., Lin, N.-W.: Lower bound cost estimation for logic programs. In: 1997 International Logic Programming Symposium, pp. 291–305. MIT Press, Cambridge, October 1997
7. Georgiou, K., Kerrison, S., Eder, K.: On the Value, Limits of Multi-level Energy Consumption Static Analysis for Deeply Embedded Single, Multi-threaded Programs. ArXiv e-prints: arXiv:1510.07095, October 2015
8. Grebenshchikov, S., Gupta, A., Lopes, N.P., Popeea, C., Rybalchenko, A.: HSF(C): a software verifier based on horn clauses. In: Flanagan, C., König, B. (eds.) TACAS 2012. LNCS, vol. 7214, pp. 549–551. Springer, Heidelberg (2012). doi:10.1007/978-3-642-28756-5_46
9. Grech, N., Georgiou, K., Pallister, J., Kerrison, S., Morse, J., Eder, K.: Static analysis of energy consumption for LLVM IR programs. In: Proceedings of the 18th International Workshop on Software and Compilers for Embedded Systems, SCOPES 2015. ACM, New York (2015)
10. Gurfinkel, A., Kahsai, T., Navas, J.A.: SeaHorn: a framework for verifying C programs (Competition Contribution). In: Baier, C., Tinelli, C. (eds.) TACAS 2015. LNCS, vol. 9035, pp. 447–450. Springer, Heidelberg (2015). doi:10.1007/978-3-662-46681-0_41
11. Henriksen, K.S., Gallagher, J.P.: Abstract interpretation of PIC programs through logic programming. In: Sixth IEEE International Workshop on Source Code Analysis and Manipulation (SCAM 2006), pp. 184–196. IEEE Computer Society (2006)
12. Hermenegildo, M., Puebla, G., Bueno, F., Lopez-Garcia, P.: Integrated program debugging, verification, and optimization using abstract interpretation (and the Ciao system preprocessor). Sci. Comput. Program. 58(1–2), 115–140 (2005)
13. Hermenegildo, M.V., Bueno, F., Carro, M., López, P., Mera, E., Morales, J., Puebla, G.: An overview of Ciao and its design philosophy. Theory Pract. Logic Program. 12(1–2), 219–252 (2012)
14. Herrmann, C., Bonenfant, A., Hammond, K., Jost, S., Loidl, H.-W., Pointon, R.: Automatic amortised worst-case execution time analysis. In: 7th International Workshop on Worst-Case Execution Time Analysis (WCET 2007), vol. 6. OASIcs. Schloss Dagstuhl-Leibniz-Zentrum fuer Informatik (2007)
15. Hojjat, H., Konečný, F., Garnier, F., Iosif, R., Kuncak, V., Rümmer, P.: A verification toolkit for numerical transition systems. In: Giannakopoulou, D., Méry, D. (eds.) FM 2012. LNCS, vol. 7436, pp. 247–251. Springer, Heidelberg (2012). doi:10.1007/978-3-642-32759-9_21
16. Jayaseelan, R., Mitra, T., Li, X.: Estimating the worst-case energy consumption of embedded software. In: IEEE Real Time Technology and Applications Symposium, pp. 81–90. IEEE Computer Society (2006)
17. Kerrison, S., Eder, K.: Energy modeling of software for a hardware multithreaded embedded microprocessor. ACM Trans. Embed. Comput. Syst. 14(3), 1–25 (2015)
18. Lafond, S., Lilius, J.: Energy consumption analysis for two embedded Java virtual machines. J. Syst. Archit. 53(5–6), 328–337 (2007)

19. Lattner, C., Adve, V.: LLVM: a compilation framework for lifelong program analysis and transformation. In: Proceeding of the 2004 International Symposium on Code Generation and Optimization (CGO), pp. 75–88. IEEE Computer Society, March 2004

20. Liqat, U., Georgiou, K., Kerrison, S., Lopez-Garcia, P., Hermenegildo, M.V., Gallagher, J.P., Eder, K., Consumption, I.E.: At Different Software Levels: ISA vs. LLVM IR. Technical report, FET 318337 ENTRA Project, Appendix D3.2.4 of Deliverable D3.2. http://entraproject.eu

21. Liqat, U., Kerrison, S., Serrano, A., Georgiou, K., Lopez-Garcia, P., Grech, N., Hermenegildo, M., Eder, K.: Energy Consumption analysis of programs based on XMOS ISA-level models. In: Gupta, G., Peña, R. (eds.) LOPSTR 2013. LNCS, vol. 8901, pp. 72–90. Springer, Switzerland (2014). doi:10.1007/978-3-319-14125-1_5

22. Méndez-Lojo, M., Navas, J., Hermenegildo, M.V.: A flexible, (C)LP-based approach to the analysis of object-oriented programs. In: King, A. (ed.) LOPSTR 2007. LNCS, vol. 4915, pp. 154–168. Springer, Heidelberg (2008). doi:10.1007/978-3-540-78769-3_11

23. Mera, E., López-García, P., Carro, M., Hermenegildo, M.: Towards execution time estimation in abstract machine-based languages. In: 10th International ACM SIGPLAN Symposium on Principles and Practice of Declarative Programming (PPDP 2008), pp. 174–184. ACM Press, July 2008

24. Navas, J., Méndez-Lojo, M., Hermenegildo, M.: Safe upper-bounds inference of energy consumption for Java bytecode applications. In: The Sixth NASA Langley Formal Methods Workshop (LFM 2008), Extended Abstract, April 2008

25. Navas, J., Mera, E., Lopez-Garcia, P., Hermenegildo, M.V.: User-definable resource bounds analysis for logic programs. In: Dahl, V., Niemelä, I. (eds.) ICLP 2007. LNCS, vol. 4670, pp. 348–363. Springer, Heidelberg (2007). doi:10.1007/978-3-540-74610-2_24

26. Pallister, J., Kerrison, S., Morse, J., Eder, K.: Data dependent energy modeling for worst case energy consumption analysis. arXiv preprint arXiv:1505.03374 (2015)

27. Rosendahl, M.: Automatic complexity analysis. In: 4th ACM Conference on Functional Programming Languages and Computer Architecture (FPCA 1989). ACM Press (1989)

28. Serrano, A., Lopez-Garcia, P., Hermenegildo, M.: Resource usage analysis of logic programs via abstract interpretation using sized types. In: Theory and Practice of Logic Programming, 30th International Conference on Logic Programming (ICLP 2014) Special Issue, vol. 14, no. 4–5, pp. 739–754 (2014)

29. Seshia, S.A., Kotker, J.: Gametime: a toolkit for timing analysis of software. In: Abdulla, P.A., Leino, K.R.M. (eds.) TACAS 2011. LNCS, vol. 6605, pp. 388–392. Springer, Heidelberg (2011). doi:10.1007/978-3-642-19835-9_34

30. Vasconcelos, P.B., Hammond, K.: Inferring Cost Equations for Recursive, Polymorphic and Higher-Order Functional Programs. In: Trinder, P., Michaelson, G.J., Peña, R. (eds.) IFL 2003. LNCS, vol. 3145, pp. 86–101. Springer, Heidelberg (2004). doi:10.1007/978-3-540-27861-0_6

31. Watt, D.: Programming XC on XMOS Devices. XMOS Limited (2009)

32. Wegbreit, B.: Mechanical program analysis. Commun. ACM **18**(9), 528–539 (1975)

33. Wilhelm, R., Engblom, J., Ermedahl, A., Holsti, N., Thesing, S., Whalley, D., Bernat, G., Ferdinand, C., Heckmann, R., Mitra, T., Mueller, F., Puaut, I., Puschner, P., Staschulat, J., Stenström, P.: The worst-case execution-time problem - overview of methods and survey of tools. ACM Trans. Embedded Comput. Syst. **7**(3), 36 (2008)

Timing Properties and Correctness
for Structured Parallel Programs
on *x86-64* Multicores

Kevin Hammond[✉], Christopher Brown, and Susmit Sarkar

School of Computer Science, University of St Andrews, Scotland, UK
{cmb21,ss265}@st-andrews.ac.uk, kevin@kevinhammond.net

Abstract. This paper determines correctness and timing properties for *structured* parallel programs on *x86-64* multicores. Multicore architectures are increasingly common, but real architectures have unpredictable timing properties, and commonly used relaxed-memory concurrency models mean that even functional correctness is not obvious. This paper takes a rigorous approach to correctness and timing properties, examining common locking protocols from first principles, and extending this through queues to structured parallel constructs. We prove functional correctness and derive simple timing models, extending these for the first time from low-level machine operations to high-level parallel patterns. Our derived high-level timing models for structured parallel programs allow us to *accurately predict* upper bounds on program execution times on *x86-64* multicores.

Keywords: Multicore · Relaxed-memory concurrency · Functional correctness · Algorithmic skeletons · Operational semantics · Timing models

1 Introduction

Multicore architectures are increasingly common, providing excellent trade-offs between performance and energy consumption. However, the actual execution behaviour of parallel programs on real multicores is still often difficult to understand. Arguing about the correctness of parallel programs is non-trivial in the presence of real-world relaxed-memory consistency models. Moreover, predicting the execution time of parallel programs is hard. Both issues derive from the same problem: *there is no well-understood correspondence between the high-level parallel primitives that programmers use and the low-level implementations of those primitives that are actually executed.* We address this problem by directly considering correctness and timing properties from basic machine-level operations all the way to high-level parallelism structures.

This paper exploits *structured parallel programming*, in the form of *algorithmic skeletons* [1,2]. Such approaches bring similar abstraction advantages to parallel program design that standard structured programming brings to sequential programming, abstracting over basic parallelism primitives, such as process creation, communication and synchronisation.

© Springer International Publishing Switzerland 2016
M. van Eekelen and U. Dal Lago (Eds.): FOPARA 2015, LNCS 9964, pp. 101–125, 2016.
DOI: 10.1007/978-3-319-46559-3_6

The main advantages to us are:

1. The programmer can think in terms of *patterns* of parallelism, rather than e.g. low-level memory operations;
2. Certain complex parallel conditions, such as race conditions and deadlocks, are eliminated *by design*, so dramatically simplifying design, implementation and testing of parallel programs; and
3. Simple, but effective, performance models are possible.

This paper exploits these advantages to produce strong and principled cost models for two fundamental parallel programming structures: *task farms* and *parallel pipelines*. Danelutto *et al.* [3] have shown that many common parallelism patterns can be expressed in terms of these two primitives. Our work thus extends straightforwardly to many other high-level patterns of parallelism.

```
int x = 0, y = 0;              extern int x, y;
...                            ...
{ // thread 1                  { // thread 2
  x = 1;                         y = 2;
  return y;                      return x;
}                              }
```

Fig. 1. Under the *x86-TSO* relaxed-memory consistency model, both threads could return 0. This is not possible under SC.

Memory Consistency. Standard correctness proofs of parallel programs usually assume memory accesses to be interleaved, so-called *Sequential Consistency* (SC) [4]. However, *x86-64* multicores, and many other recent architectures (e.g. ARM and IBM Power), do not comply with this assumption, making such reasoning invalid. Consider, for example, the program fragment in Fig. 1. Under the *Total Store Order* relaxed-memory consistency model that is used by *x86-64* multicores (*x86-TSO*), both threads could return 0, a result that is impossible using the simpler SC model.

Novel Contributions. While there has been significant previous work on algorithmic skeletons, this paper represents the first attempt, of which we are aware, to establish cost models from first principles for widely-used multicore hardware. In addition, this paper makes the following specific novel contributions:

1. It describes an operational semantics for structured parallel programs, that is used to derive, from first principles, a compositional cost model for any combination of the farm and pipeline algorithmic skeletons;
2. It gives a simple operational proof of the (partial) correctness of a widely-used *spin-lock* protocol using the actual relaxed-memory concurrency semantics used by *x86-64* multicore machines; the proof composes in a straightforward

fashion to show that the farm and pipeline algorithmic skeletons are free from deadlock, both singly and in any combination, when implemented by queues using that locking protocol; and

3. It validates the cost model against some example parallel programs, giving accurate predictions of lower-bound speedups; our predictions are typically within 7 % and in all cases within 30 % of the actual speedup.

A key aspect of our approach is that we model the actual relaxed-memory model used by real *x86-64* multicore systems, exploiting recent work on the *x86-TSO* model [5,6], which gives a precise account of the observable behaviour of *x86-64* multiprocessors in terms of an idealised abstract machine. We are thus able, for the first time, to provide realistic cost models for programs that are executed in parallel on such architectures. The spin-lock protocol that we use has been shown to be the most efficient protocol for implementing simple locks on *x86-64* architectures [5]. It is widely used to implement real parallel programs. Our correctness proof is the first direct operational proof that this protocol is correctly implemented by the underlying hardware through its relaxed-memory access protocol. Finally, the queue protocol shown here has many uses beyond algorithmic skeletons, and is representative of many widely-used higher-level synchronisation protocols. The correctness proof shown here is likewise the first proof that such a protocol is correctly implemented by the underlying multicore hardware.

2 Pipeline and Farm Skeletons

In this paper, we consider two fundamental skeletons: 2-stage *pipelines* (Fig. 2) and *task farms* (Fig. 3). We use a streaming implementation that links skeletons using (unbounded) queues. This allows skeletons to be easily composed, and also allows arbitrarily nested parallel structures to be built from the basic skeletons. Parallel pipelines $(f \mid g)$ represent the composition of two operations f and g, streamed over a sequence of inputs $t_0, t_1, \ldots$, with f and g possibly executed

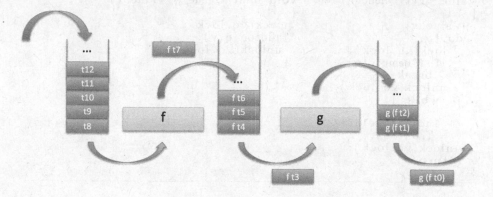

Fig. 2. Two-stage parallel pipeline: $f \mid g$

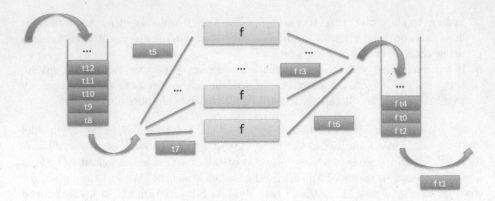

Fig. 3. Task Farm: $\Phi(f)$

in parallel. The result of the pipeline is the stream g $(f\ t_0), g$ $(f\ t_1), \ldots$ Multi-stage pipelines can be built by composing multiple two-stage pipelines and then merging the input/output queues. *Task farms* $(\Phi(f))$ apply the same operation f to each of the inputs in a stream. A fixed number of worker instances is created, which each apply f in parallel to a subset of the inputs. The result of applying a farm to a stream of inputs $t_0, t_1, \ldots$ is then the *bag* $\{f\ t_0, f\ t_1, \ldots\}$, where the results may be produced in an arbitrary order. This *non-deterministic* definition of a task farm allows an efficient parallel implementation, where each of the workers is mapped to a different processing agent. As each worker produces its result, it is placed in the output queue, and the worker takes the next input (if any) from the input queue. Farms can be nested to an arbitrary depth by replacing the operation f by a farm and linking the corresponding input and output queues. It is also possible to embed pipelines within farms or *vice-versa*, so yielding arbitrarily complex parallel systems.

```
Value qget(Queue q)              void qput(Queue q,Value v)
{                                {
  Value v;                         lock(&q.lock);
  do {                             addtoq(q,v);
    lock(&q.lock);                 unlock(&q.lock);
    if (!qempty(q))              }
      break;
    unlock(&q.lock);
  } while (1);

  /* lock is held */
  v = front(q);
  unlock(&q.lock);
  return(v);
}
```

Fig. 4. Simple queue implementation in C.

2.1   Simple Queue Implementation Using Locks

The pipeline and farm skeletons use queues to manage the input/output streams. Figure 4 shows how queues can be implemented using `lock` and `unlock` primitives, plus operations to remove the first element from the queue (`front`), add a new element to the queue (`addtoq`), and check whether the queue is empty (`qempty`). Queues are implemented using the abstract type `Queue`, containing values of type `Value`. The `qget` operation returns the first element from the queue, spinning if the queue is empty. This implements a blocking read operation. The corresponding `qput` operation adds a value to the end of the queue, locking and unlocking as necessary. When used as a pair, `qget`/`qput` implement a synchronisation operation between two parallel threads. Figure 5 gives a spin-lock implementation of `lock`/`unlock` in C and *x86-64* assembler, assuming an atomic exchange primitive, `exchange`. Spin-locks are widely used in parallel systems where there are low contention rates, since they are simple to implement and the costs of acquiring and releasing the locks are very low. In fact, until recently, the Linux kernel used almost identical code. For fairness reasons, it now uses a slightly modified version (*ticketed spin-locks*). Here, `lockcell` is a variable that contains the lock. If it has the value 1, then the lock has been acquired by some thread; if it has the value 0, then no thread has acquired the lock. Acquiring the lock using the `lock` routine involves reading the value of `lockcell`, and exchanging it with the locked value, 1. If the lock has already been acquired by another thread, the process is repeated until the lock can be acquired (i.e. the previous value of `lockcell` was 0). Releasing the lock using the `unlock` routine simply involves setting `lockcell` to 0. Note that the unlock need not be an atomic memory operation.

3   Key Hardware Characteristics

In what follows, we will consider memory accesses to be classified into *Reads* from and *Writes* to specific memory locations, plus *Fences* and atomic *Exchanges*. Intel's *x86-64* instruction set also provides some refined versions of these basic operations [7]. *Sequential consistency* (SC) memory models [4] ensure that memory accesses from multiple threads are carried out in an order that is consistent with some valid set of memory accesses by a fully sequential processor. That is, memory accesses from different threads are interleaved so that there is effectively a single thread of memory accesses. Recall the simple example from Fig. 1. Here, depending on the exact timing of *Reads* and *Writes* on x and y, thread 1 could return either 1 or 0 and thread 2 could return either 2 or 0. It is not possible for both threads to return 0, however, since one of the two *Writes* to x or y must happen last. While SC is effective on uni-processor systems, enforcing an SC memory model on a multicore system can carry significant performance penalties. For example, all caches and other memory hardware must be synchronised in order to avoid inconsistent results. Since such a strong model is not always required, multicore hardware vendors generally support weaker

```
void lock( volatile char *lockcell ) {
    char old_value ;
    do {
        old_value = exchange(lockcell , 1);
    } while ( 1 == old_value ) ;
}

void unlock( volatile char *lockcell ) {
    *lockcell = 0 ;
}
```

```
; Assume EBX contains address of lock cell
_lock:
    mov  eax, 1      ; Set EAX register to 1 (locked)
    xchg eax, [ebx]  ; Exchange EAX and lock cell
    test eax, eax    ; Test if cell is locked
    jnz  _lock       ; Retry the lock if so
    ...              ; Lock held here


_unlock:
    mov  eax, 0      ; Set EAX register to 0 (unlocked)
    mov  [ebx], eax  ; Release the lock
```

Fig. 5. Above, simple spin-lock implementation in pseudo-C using an atomic **exchange** operation on the **lockcell** memory location; 1 indicates the lock is acquired; 0 is used to release the lock. Below, corresponding *x86-64* instructions (ignoring function prelude/postlude). The **xchg** instruction in _lock acquires the lock, which is released in in _unlock using a normal memory write. Note that on *x86-64*, a **xchg** implicitly behaves as a full memory barrier.

consistency models that offer higher performance. Modern *x86-64*-class microprocessors use Total Store Ordering (*x86-TSO*) [5,6]. *x86-TSO* guarantees that the order in which *Write* instructions for a given processor appear in (shared) memory is identical to the sequence in which the processor issued the *Writes*. *x86-TSO* can be implemented by providing each hardware thread of execution with a private FIFO write-buffer. This is an abstract machine implementation: a more realistic hardware implementation is discussed further below. *Writes* are stored temporarily in this buffer prior to being actioned by the main memory. *Reads* from the local processor (only) can access this write buffer, if necessary, as an intermediate step between reading from the private and shared caches. *Writes* recorded in the write buffer will be used in preference to any corresponding values in the shared cache. After an unpredictable, but finite, time, each *Write* is flushed from the buffer, and so becomes visible to all processors. In this way, SC is enforced for a single processor, *but not for all processors in a multi-processor system.* In the *x86-TSO* model, explicit *memory fence* (**mfence**) or *atomic exchange* (**xchg**) instructions are needed to enforce consistency across multiple processors. When a Fence/Exchange instruction is encountered by a processor, all of its outstanding *Reads* and *Writes* are executed immediately.

This provides strong local temporal guarantees: all local accesses that appeared in the instruction stream prior to a *Fence/Exchange* will be executed *before* any accesses that appeared after the *Fence/Exchange*. It also provides strong global memory guarantees: all memory locations (and any cached copies) will be consistent with the memory state immediately following the *Fence/Exchange*.

Definition 1. *Thread. A* Thread *is an ordered sequence of memory accesses, where memory accesses comprise* Reads, Writes, Fences *and* Exchanges *to specific memory locations. A* Read *takes a memory location and returns a value. A* Write *takes a memory location and a value and has no result. A* Fence *has no parameters or result. An* Exchange *is treated as a simultaneous and indivisible* Read, Write *and* Fence. *It takes a memory location and a value and returns a (possibly different) value.*

Definition 2. *Write buffer. A* Write buffer *is associated with each thread, and contains an ordered sequence of* Writes, *each associating a value with an address. Each* Write *is recorded in the corresponding write buffer immediately it is executed by any* Thread.

Definition 3. *Memory. A* Memory *maps addresses to values.*

Definition 4. *Core. A* Core *comprises a set of* Threads.

Definition 5. *Multicore. A* Multicore *comprises a set of* Cores *plus a single shared* Memory[1].

Definition 6. *Execution Order. The* execution order *is a linear trace of transitions made by the labelled transition system called the x86-TSO machine [6].*

Hardware Correspondence. The abstract model above talks about "threads" of execution. In current multiprocessor implementations, these are grouped together in various ways, each of which has varying implications for scheduling and timing. Firstly, *user-level threads* are mapped by the *runtime system* to *hardware-level threads*. In this paper, we ignore this scheduling cost. That is, we consider only a non-preemptive model with no explicit thread *yields* to other threads or the runtime system. Taking such yields into account would introduce interference with other processes and threads running on the system. The model above speaks of such hardware threads when mentioning a thread of execution. Secondly, each hardware thread is mapped to specific hardware resources on a core. The mapping may be one-to-one, or many-to-one (*hyperthreading*, also known as *simultaneous multi-threading*). Each hyperthread generally has exclusive use of registers and a load-store reordering buffer. The load-store reordering buffer maintains metadata to ensure observable FIFO buffering for the stores, and also to ensure that loads appear ordered to the programmer, even though

---

[1] We will ignore shared cache here, since it does not have a significant impact on the proofs.

aggressive implementations can and do perform out-of-order operations, e.g. satisfaction of read requests. Third, a set of cores (typically 2 to 4) are collected in one CPU. Typically, the first levels of cache (L1 and L2 on the Intel Core-i7) are private to the core, while higher levels (L3 on the Intel Core-i7) are shared between all the cores on one CPU. Finally, multiple CPUs may be connected in one system, all sharing a common memory. The caches communicate with each other and with memory to maintain "cache coherence", that is, a clear notion of order of update operations (stores) to a location. This is generally managed at the cache-line granularity (64 on the Intel Core-i7). To be clear, the functional model above speaks of store buffers. These are implemented by so-called Memory Ordering Buffers (MOBs), which internally do out-of-order actions. The flushing of buffers then corresponds to the point when non-local threads (on the same core or otherwise) can see those stores. Further buffering occurs between various levels of the cache. However, in all cases the cache maintains a coherent view, so that if the store is visible to at least one hardware thread other than the one executing the store, then it is visible to all of them.

X86-64 Cache Protocol. Both the Intel and AMD implementations of the *x86* cache protocol are variations of the classic MESI protocol [8]. Conceptually, there can be four states for each location in a cache (managed on a cache-line granularity): **M**odified (this cache holds an exclusive copy and memory has a stale copy), **E**xclusive (this cache holds an exclusive copy, the memory copy is also valid), **S**hared (this cache holds a copy, but other caches possibly hold copies as well), and **I**nvalid (this cache does not hold a current copy for this location). While the cache behaviour does not directly impact the functional correctness proof, it does have implications for the cost models we develop.

4 Correctness and Progress Properties

This section considers the functional correctness of the spin-lock, queue and pipeline/task farm implementations described above, working from first principles in terms of the basic *x86-64* memory operations and *x86-TSO* consistency model. We first consider a key ordering relation, *coherence order*, then sketch soundness proofs for the spin-lock and queue implementations, and finally build on these to sketch soundness proofs for the skeleton implementations.

4.1 Coherence Order

A derived relation called *coherence order* naturally emerges from the *x86-TSO* memory consistency model described above. *Coherence order* is a total, linear ordering of *Writes* to a given memory location, organised in the order that the *Writes* affect the memory location (are *flushed* from their local buffers).

Definition 7. *Coherence Order. Given a set of* write buffers, WB_i *containing tuples* $\langle t, m, v \rangle$, *where t is the time that the* Write *is flushed from* WB_i, *m is the* *memory location, and v is the new value to be written to that location, then the* coherence order *for some memory location m is defined to be the sequence*

$$CO(m) \ linear \ order \ over \ \{\langle t_j, m, v_j \rangle \mid \langle t_j, m, v_j \rangle \in \bigcup_{i=1}^{n} WB_i\}$$

$$s.t. \quad \forall j, k. \langle t_j, m_j, v_j \rangle <_{CO(m)} \langle t_k, m_k, v_k \rangle \implies t_k > t_j.$$

Note that under this definition, it is not possible for two *Writes* to the same memory location to occur at the same time. *Writes* to different memory locations may, however, occur at the same time. This is consistent with the restrictions of physical memory. It is easy to show from this definition that no *Thread* can read values out of coherence order.

Lemma 1. *Read Coherence Order. No Thread can read values out of coherence order.*

Proof Sketch: Suppose that two *Reads* r_1 and r_2 in the same *Thread*, T, from the same memory location, m, occur in order, but that they return different values. Let us call the corresponding *Writes* w_1 and w_2 ($w_1 \neq w_2$). We do a case analysis depending on whether w_2 is read by r_2 from the local write buffer or from memory. Suppose that w_2 is in the *write buffer*. Then w_2 cannot have been the last write in the write buffer at the time of the read r_1, and thus w_2 must be flushed at some later time, and definitely later than w_1. Suppose that, instead, w_2 is taken from memory. Then the local write buffer must be empty for that location. Now, either r_1 read w_1 from the write buffer, and therefore w_1 must have been flushed (before w_2), or it read it from memory, and again w_2 is in *Coherence order* before w_1. □

4.2 Functional Correctness of the Spin-Lock Implementation

The spin-lock implementation needs to enforce two key properties: (i) that at most one thread at a time possesses the lock; and (ii) that all *Writes* that are made while the lock is held are always visible to any subsequent thread that acquires the lock. We show this by using the memory properties defined in the previous section, together with the code for the spin-lock implementation of `lock` and `unlock` that was given in Fig. 5.

Lock Acquisition and Release. We assume that the only *Reads/Writes* to `lockcell` are made by the `lock` and `unlock` functions. We also assume that `unlock` is called only when safe, i.e. when the calling *Thread* possesses the lock, and that `lock` is called only when the calling *Thread* does not possess the lock. It follows that the only values that `lockcell` can contain are 0 (the unlocked value) and 1 (the locked value).

Theorem 1. *Lock acquisition and release. Under the assumptions of the previous paragraph,*

– *following a call to* `lock`, *the calling* Thread *will possess the lock, and no other* Thread *will possess the lock; and*
– *following a call to* `unlock`, *the calling* Thread *will no longer possess the lock until it has successfully called* `lock` *again.*

Proof Sketch: Both of the above parts can be proved simultaneously by induction on the `lock` and `unlock` calls in an execution trace, and by case analysis of the write buffer state at every lock. The key operation is the `lock` function, which is called when acquiring the lock. We return from this function exactly when the internal loop exits, that is when `old_value` is not 1. This means that `old_value` is 0, or in other words, that `lockcell` held 0 in the final iteration of the loop and is now 1. Since the only way to update `lockcell` with a locked (1) value is via an Exchange, this will always appear directly in memory, and never stay in the write buffer. In contrast, an unlocked (0) value can stay in the buffers, since this is done by a simple *Write*. Since only the *Thread* that has successfully acquired the lock ever calls the `unlock` function, and since it does this precisely once for each lock acquisition, there will only ever be at most one unlock value (0) for `lockcell` in any of the write buffers. There are now two cases to consider.

Case 1: Consider first the case when no write buffer contains an unlock value. If `lockcell` holds 1, then the lock is already held. No lock acquisition can succeed until `lockcell` becomes 0. Conversely, if `lockcell` holds 0, the lock is free. Any isolated lock acquisition will successfully acquire the lock, but because of the atomic nature of the *Exchange*, if there are multiple simultaneous acquisition attempts then only one of these can succeed. Furthermore, because of coherence on `lockcell`, no subset of threads can disagree on which lock acquisition succeeded.

Case 2: The second case occurs when some *Thread T*'s write buffer contains an unlock value. It follows that `lockcell` must still be locked, and therefore no *Thread* can acquire the lock. However, thread T has released the lock. Now either the unlock value must eventually flush by itself, and we will then be in Case 1 above, or T can attempt to re-acquire the lock, in which case the first action of the *Exchange* will be to flush the write buffer.   □

We thus obtain the first key property of the lock implementation, that the lock, from an initially unlocked position, flips between the locked and unlocked states, and because of coherence, no set of threads can disagree about which thread was responsible for each lock and unlock operation in that sequence. In other words, two distinct threads never think that they hold the lock at the same time.

Synchronisation. We now turn to the second key property, that the lock acquisition and release operations collectively provide synchronisation. More precisely, following the acquisition of a lock, a thread should have an identical view of the shared memory to that of the unlocking thread at the immediately preceding lock release. This is not an immediate result, since the two threads may be different, and the view of the shared memory is mediated by the write buffers of the cores that are involved.

Theorem 2. *Synchronisation. Assuming shared-memory accesses are only performed by threads holding a lock, all* Writes *made by a* Thread *between a successful call to* lock *and the following call to* unlock *are visible to the next* Thread *that acquires the lock.*

Proof Sketch: We observe that each write buffer is emptied in a FIFO manner, and that each *Thread* has its own write buffer. *Writes* to shared memory locations within the critical section controlled by the lock are initially buffered in the write buffer for the *Thread* that acquired the lock. The lock release is itself a *Write*, that is placed in the buffer *after* all these *Writes*. We proved above that a new lock is only acquired when the unlocking *Write* is flushed from the buffer. Since the write buffer is FIFO, this implies that all the preceding *Writes* in the critical section must have already been flushed to the *Memory*. Now turning to the lock acquisition, since a lock acquisition involves an *Exchange*, which completely flushes a *Thread*'s write buffer, any *Read* to the shared locations (in case of multiple accesses to one location, the first such) by the acquiring *Thread* must read its value from memory. It follows that synchronisation is obtained. □

4.3 Progress of the Spin-Lock Implementation

When acquiring a lock, the *Thread* calling the lock function will loop until it succeeds. Conversely, the unlock function will always succeed. Progress of the system depends on two assumptions about the memory system. Firstly, if there are multiple contending *Exchanges*, exactly one will succeed. Secondly, all write buffers will eventually flush their contents to *Memory*. The first assumption is required to allow progress in the case of contending lock-acquires when the lock is free. The second assumption is required to allow progress when one thread has released a lock and other threads are waiting to acquire the lock. Note that if the same thread later wants to acquire the lock, the *Exchange* within lock will automatically flush the buffers. This is the progress condition for hardware exchanges that was discussed above.

4.4 Functional Correctness of the Queue Implementation

Recall that the pseudo-code for the queue implementation was shown in Fig. 4. We will assume that the queue is only touched by the qget and qput functions that are defined here, and the underlying lock location is not directly accessed by other code. We first verify that the queue implementation validates the assumptions made in the proof of the spin-lock implementation. First, since the only code that affects lockcell are the calls to the lock and unlock functions, it is easy to see no other code touches lockcell. Furthermore, on every control flow path, the unlock function is only called after having acquired the lock, it is not called more than once, and it is definitely called before either qget or qput returns. Moreover, any access to the queue and its fields occurs in program-order between a `lock` and an `unlock` from the same thread. Now we argue that the queue properties are ensured by the operations. There are three key properties:

1. The `qget` operation returns with a value that was previously placed in the queue.
2. A value that is placed once in the queue is never removed twice.
3. **FIFO** order is maintained for the queue.

For (1), the loop in the `qget` function can only be exited by the **break** statement. This means that when the loop exits, we know that the queue is non-empty, and furthermore, that the *Thread* that called `qget` holds the lock. Thus no other thread can remove elements from the queue, and it follows that the `front` operation will be able to find and remove at least one value. For (2), since the `front` operation completes before the lock is released, a value cannot be removed more than once. For (3), since only one thread can execute either `qget` or `qput` at a time, the queue follows sequential semantics. Therefore values are removed in the order that they were put into the queue.

4.5   Progress of the Queue Implementation

Since there is only one lock per queue, there is no possibility of deadlock. Progress depends on the progress properties of the underlying spin-lock implementation, in that if there are multiple contending `qget` and `qput` calls, at least one must succeed. Progress also depend on the fairness of the spin-lock implementation, which is an additional assumption. To see why, consider the case when the queue is empty, and there are one or more `qget` operations contending with a `qput` operation. An unfair implementation could allow a `qget` operation to succeed in acquiring the lock, notice that the queue is empty, release the lock, and immediately acquire the lock again without allowing the `qput` operation to acquire the lock. In this case, the `qput` would be starved, and the system as a whole would be prevented from making progress.

4.6   Functional Correctness of the Farm/Pipeline Skeletons

As defined here, the pipeline and farm skeletons are both streaming operations, with workers applying functional operations to inputs to produce some output. The requirement is that all inputs that are placed in the input streams are processed to produce a corresponding output. Moreover, each input in a stream is processed precisely once. Each worker is associated with an input queue $Q_i = x_1, x_2, \ldots, x_n$ and an output queue Q_o, containing tasks and results, respectively. Each queue is guarded by its own lock, and is shared with one or more other workers. A worker obtains an input task from its input queue, applies its functional operation, and then places the result of the task on the output queue. It is easy to see that the assumptions that we need to ensure the correctness of the queue implementation (the queue is touched only by the qget and qput methods) are valid for both of the queues. Furthermore, any communication between workers occurs only through these queues.

Let us consider a parallel pipeline $f \mid g$. From the queue properties, by induction on the initial sequence in the input queue $x_1, x_2, \ldots, x_n$, and assuming that f is finite, we can see that worker f will produce results in the sequence $f(x_1), f(x_2), \ldots, f(x_n)$, and that all inputs will produce a corresponding output. Again, by the queue properties and by induction on this sequence, the next stage will produce the sequence $g(f(x_1)), g(f(x_2)), \ldots, g(f(x_n))$. That is, $\forall j, 0 < j \leq n.Q_o[j] = g(fQ_i[j])$, and so the output queues Q_o represents a map of $g \cdot f$ over the elements in the input queues Q_i.

Now let us consider a task farm $\Phi(f)$. Suppose that the initial sequence in the input queue is $x_1, x_2, \ldots, x_n$. By the queue properties, each element is removed exactly once, and assuming that f is finite, all workers will complete and put results $f(x_i)$ on the output queue. However, without further limitations, the order of the results in the output queue will now depend on the scheduling of each of the worker threads. Unlike the strong ordering for the pipeline, here we can only say that the output queue will contain $f(x_1), f(x_2), \ldots, f(x_n)$ in some permutation. It is not possible to make a strong statement about result ordering without further knowledge of the thread scheduler.

4.7 Progress of the Farm and Pipeline Skeletons

Since both the farm and the pipeline skeletons do not contain any high-level cycles, they satisfy progress. The only possible violations of progress are then when the queue operations are called, with either blocking or deadlocks. We have proved that the queue operations satisfy progress above (assuming a fair spin-lock implementation), and thus proving progress under the same assumption here is almost trivial. The ease of this argument emphasises the advantages of taking a structured view.

5 Timing Models for *x86-64* Multicores

Our overall objective is to obtain good timing predictions for parallel code running on *x86-64* multicores, based on a rigorous understanding of the underlying relaxed memory model. As mentioned in the hardware correspondence section, we consider a non-preemptive model. This means that our timing models directly apply when parallelism is less than or equal to the number of cores. Alternatively, if parallelism is greater than the number of cores, stronger fairness assumptions on the scheduler will have to be made to extend our timing models. In order to construct our timing cost models, we will use *traces* to describe the operation of each thread in the system, and then abstract over these to determine the overall timing behaviour of the system. A *trace* describes an actual execution of a thread in terms of the observable primitive operations that it performs. Since we are not interested in the values that our system produces, but only in the time that it takes to execute, we abstract over the actual computations that a thread performs between each memory access, using the abstract *Compute* operation to capture the time taken by actual computations. This paper considers only the costs associated with memory accesses. In the actual hardware, memory access costs depend on a variety of factors such as the presence or absence of a location in cache, the interconnect topology, the relative speeds of the various cache levels of the cache, etc. There are, thus, varying levels of realism that can be included in the cost model. We begin with a relatively naïve model, which just considers buffer sizes. As we will see, even a simple model like this already captures most of the important timing effects that we are interested in.

5.1   Simple Average Timing Model

Our first timing model estimates execution costs by assigning an average time to each kind of access. Our parameters are:

T_{Read} the (average) time to read a location
T_{Write} the (average) time to write a location
$T_{Exchange}$ the (average) time to exchange a location

We build up our model in stages, starting with the spin-lock implementation from Fig. 5, then considering the queue implementation from Fig. 4, and finally extending our model to the high-level farm and pipeline skeletons.

Spin-lock Timings. The cost of an `unlock` operation is just T_{Write}, while that of a `lock` operation is $N \cdot T_{Exchange}$, where N is the number of times that a thread spins before acquiring the lock. Since each lock can only be held by one thread at a time, if a thread attempts to acquire a lock while it is held by another thread, it will spin uselessly. Once the lock is released, if t threads are all trying to acquire the lock, only one will succeed. Assuming that the hardware allows precisely one thread to exchange successfully, it follows that it will take $t \cdot T_{Exchange}$ time before the lock is acquired by some thread.

Queue Timings. Suppose that there are n threads accessing the queue. In the worst case, they will all contend with each other. By the argument above, a `qput` operation therefore succeeds in time: $T_{qput} = n \cdot T_{Exchange} + T_{Write} + T_{Write}$, i.e., $n \cdot T_{Exchange}$ for the lock; one T_{Write} when adding to the queue; and another T_{Write} for the unlock. Similarly, *provided that the queue is not empty*, a `qget` operation succeeds in time $T_{qget} = n \cdot T_{Exchange} + T_{Read} + 2T_{Write}$ i.e., $n \cdot T_{Exchange}$ for the lock; one T_{Read} when reading the head of the queue; and two T_{Write}, one for writing the queue head and one for the unlock.

Farm and Pipeline Skeleton Timings. For the *farm* skeleton, suppose there are n worker threads. The work done by each thread is a `qget` followed by the actual computation and finally a `qput` operation. Each of the queue operations has contention $n + 1$. Thus, each thread takes time $T_{qget} + T_f + T_{qput}$ which simplifies to $2 \cdot (n + 1) \cdot T_{Exchange} + 4 \cdot T_{Write} + T_{Read} + T_f$. For the two-stage *pipeline* skeleton, there are two cases depending on which of the two stages dominates the time taken (the other being idle). If the first stage dominates (qualitatively, the second stage has insufficient work), then the first queue has contention $|f|+1$, the middle queue has contention $|f|+1$, and the second queue has contention 1. The first stage thus takes time $T_{\text{qget}}+T_f+T_{\text{qput}}$ which simplifies to $2 \cdot (|f|+1) \cdot T_{Exchange} + 5 \cdot T_{Write} + T_f$. while the second stage takes time T_g. The total time is the sum of the times for the two stages $2 \cdot (|f|+1) \cdot T_{Exchange} + 5 \cdot T_{Write} + T_f + T_g$. Conversely, if the second stage dominates (qualitatively, the second stage is saturated), then the first queue has contention $|f|+1$, the middle queue has contention $|f| + |g| + 1$, and the second queue has contention 1. The first stage takes time T_f, and the second takes $T_{\text{qget}} + T_g + T_{\text{qput}}$ which simplifies to $(2 \cdot |g| + |f| + 1) \cdot T_{Exchange} + 5 \cdot T_{Write} + T_g$ Again, the total time is the sum of the times for the two stages $T_f + (2 \cdot |g| + |f| + 1) \cdot T_{Exchange} + 5 \cdot T_{Write} + T_g$.

5.2   Including Store Buffer Flushing

A more realistic model would take into account that an exchange operation has two components: (i) flushing the local store buffer, T_{Fl}; and (ii) the actual exchange operation T_{JustX}. Naturally, the first component depends on the number of entries in the local buffer that are waiting to be flushed, b. It follows that $T_{Exchange} = b \cdot T_{Fl} + T_{JustX}$.

Spin-lock Timings. Notice that the exchange operation is only used in the context of a tight loop in the lock operation. The first time, the store buffer must be flushed, but each successive time, the store buffer is already empty. The price of a flush is thus paid only once. Assuming that t threads contend for a lock as before, and with the same assumption as before that a single exchange succeeds, then the cost of a `lock` operation is $b \cdot T_{Fl} + t \cdot T_{JustX}$. Here, b is the (average) number of writes waiting in the buffers, which is between 0 and the size of the store buffer, B. The `unlock` operation has the same cost as in our first model.

Queue Timings. The `qput` operation now takes time: $b \cdot n \cdot T_{Fl} + n \cdot T_{JustX} + T_{Write} + T_{Write}$, while the `qget` operation takes time: $b \cdot n \cdot T_{Fl} + n \cdot T_{JustX} + T_{Read} + 2 \cdot T_{Write}$.

Farm and Pipeline Skeleton Timings. A one-stage farm takes: $2 \cdot (n + 1) \cdot b \cdot T_{Fl} + (n+1) \cdot T_{JustX} + 4T_{Write} + T_{Read} + T_f$. For a two-stage pipeline, the same analysis as before applies. If the first-stage dominates, the time is: $2 \cdot (|f|+1) \cdot b \cdot T_{Fl} + (2 \cdot |f| + |g| + 1) \cdot T_{JustX} + 5 \cdot T_{Write} + T_f$ and if the second-stage dominates, the time is: $(2 \cdot |g| + |f| + 1) \cdot b \cdot T_{Fl} + (2 \cdot |f| + |g| + 1) \cdot T_{JustX} + 5 \cdot T_{Write} + T_g$.

5.3   Predicted Speedups

Speedup for Farms. We now use the cost model above to predict execution speedups for some example skeletons. First consider a simple one-stage farm with n threads uniformly doing T_f work. The computation takes time $n \cdot T_f$, which we can consider to be constant when calculating speedup. The total cost is as calculated for farms in Sect. 5.2. Assuming that T_f is sufficiently large to dominate the time of a single *Read*, *Write*, or pure *Exchange*, the predicted speedup is approximately: $\text{Speedup}_\Phi \simeq \frac{n \cdot W}{1 + c \cdot n^2}$, where W is a constant that depends on the total cost of computation, and c is a constant that depends on the number of times that the lock is taken. Both these constants can be determined by instrumenting the code when it is run sequentially on just one core.

Speedup for Two-Stage Pipelines. Now consider a two-stage pipeline with f first-stage workers and g second-stage workers, each uniformly doing work T_f and T_g, respectively. Since the total work is constant, if the first stage is fully occupied then the predicted speedup will be: $\text{Speedup}_{f|g} \simeq \frac{f \cdot W_1}{1 + c_1 \cdot f}$ (*eqn.* 1), where W_1 is a constant that depends on the total cost of computation done in stage one (and can be approximated by the cost of a single thread in the first and a single thread in the second stages), and c_1 is a constant that depends on the number of times that the lock on the first queue is taken. If the second stage is saturated, then the predicted speedup will be: $\text{Speedup}_{f|g} \simeq \frac{f \cdot W_2}{1 + c_2 \cdot (f+g)}$ (*eqn.* 2), where again W_2 is a constant depending on the total cost of computation, and c_2 is a constant depending on the number of times the lock on the second queue is taken. Constant W_2 will be T_g / T_f times the constant W_1 above (and thus constant for a particular application). The stage that dominates depends precisely on which of the values from Eq. (1) or (2) is lower, i.e. precisely when the second stage becomes saturated. The final speedup is thus: $\text{Speedup}_{f|g} \simeq \min \left(\frac{f \cdot W_1}{1 + c_1 \cdot f}, \frac{f \cdot W_2}{1 + c_2 \cdot (f+g)} \right)$.

6   Experimental Validation

We evaluate our cost models against real execution costs using a number of different benchmarks, running on three different *x86-64* multicores. The main system that we use (*titanic*) is a 2.4 GHz 24-core, AMD Opteron 6176 architecture, running Centos Linux 2.6.18–274.el5, and gcc 4.4.6. We also use *ladybank*, a 2.93 GHz 8-core (plus hyperthreading) Intel Xeon X5570 architecture, running GNU/Linux 2.6.32–358.6.2.el6, and *lovelace*, a 2.3 GHz 64-core, AMD Opteron 6376 architecture, running GNU/Linux 2.6.32–279.22.1.el6, which we used to test scalability for a limited set of experiments. All the results reported here are averages of 10 runs on an idle machine (but not in single-user mode). All speedups are absolute results against the original sequential versions.

6.1   Image Convolution

Image convolution is widely used in image processing applications such as blurring, smoothing or edge detection. The convolution algorithm is a composition of two functions $r \circ p$, where r reads in an image from a file and p processes the image by applying a filter. This process is typically applied to a stream of input images, and produces a stream of output images. For each pixel in the input

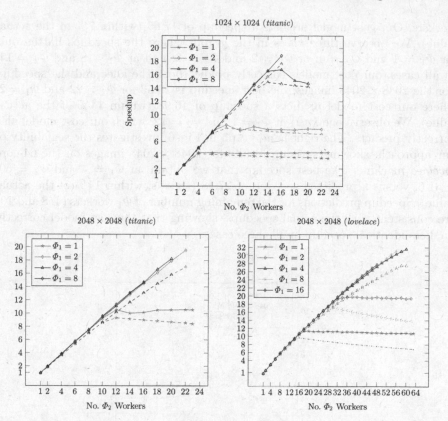

Fig. 6. Image Convolution ($\Phi_1(r) \mid \Phi_2(p)$) on *titanic* and *lovelace*. Dashed/lighter lines are predictions. Note that the number of $\Phi_1 + \Phi_2$/workers $\leq$ total hardware threads (24 for *titanic*/64 for *lovelace*).

image, the filtering stage consists of computing the scalar product of the filter and the window surrounding the pixel:

$$output\_pixel(i,j) = \sum_m \sum_n input\_pixel(i-n, j-m) \times filter\_weight(n,m)$$

For our benchmark tests, we parallelise the convolution using a two-stage pipeline, where each stage is a farm: $\Phi_1(r) \mid \Phi_2(p)$. Speedup results for two different image sizes on the 24-core *titanic* machine are shown on the left and centre of Fig. 6, with each worker in either farm allocated to its own core. Speedups are recorded against the sequential version of the algorithm. The solid lines show the actual execution speedups and the dashed lines show the speedup predictions. The x axis is the number of Φ_2 workers and each line corresponds to a fixed number of Φ_1 workers. In all cases, the predicted speedups closely match the actual results, giving a correct lower bound prediction of the actual speedup. For the 1024×1024 images, the best speedup is 18.90 for $\Phi_1 = 8$ and $\Phi_2 = 16$

workers. Our cost model predicts a speedup of 17.65 (within 7% of the actual value). We observe three *knees* in the graph, where the speedups flatten out: for $\Phi_1 = 1$ and $\Phi_2 = 4$; or $\Phi_1 = 2$ and $\Phi_2 = 6$; and at $\Phi_1 = 8$ and $\Phi_2 = 14$. In all cases, our cost models correctly predict both the knee and the speedup. For the 2048 × 2048 images, the best speedup is 19.53 for $\Phi_2 = 22$ and $\Phi_1 = 2$, where our cost model predicts a speedup of 16.98 (within 14% of the actual value). We observe one *knee*, at $\Phi_1 = 1$ and $\Phi_2 = 12$, which our cost model also correctly predicts. The right-hand graph in Fig. 6 investigates the scalability of our approach, showing speedup results for 2048 × 2048 images on the 64-core *lovelace* machine. The best speedup that we obtain, at $\Phi_1 = 4$ and $\Phi_2 = 60$, is 31.6, versus a predicted speedup of 27.48 (that is, within 14% of the actual value). Speedup predictions for the remaining number of Φ_1 workers (4, 8 and 10) are consistent with the actual speedups, showing that the cost model correctly predicts identical speedup in all three cases.

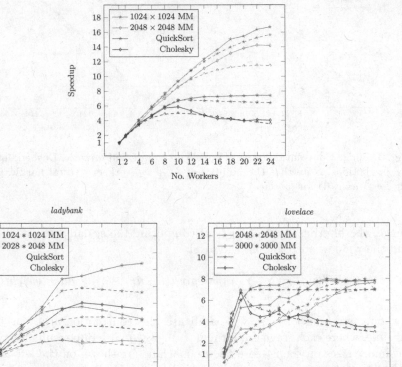

Fig. 7. Speedups for Cholesky Decomposition, Matrix Multiplication and QuickSort, using single task farms, on *titanic*, *ladybank* and *lovelace*. Dashed lines are predictions.

6.2 Cholesky Decomposition

Cholesky Decomposition is used in linear algebra, comprising the decomposition of a Hermitian, positive-definite matrix into the product of a lower triangular matrix and its conjugate transpose. Our implementation uses a task farm, Φ_1, to model the decomposition stage. Speedup results for a 1024×1024 matrix are shown in Fig. 7. The top set of results shows speedups for the 24-core *titanic* machine (an AMD architecture) and the bottom set of results shows the corresponding speedups on the 8-core (*hyperthreaded*) *ladybank* machine (an Intel Xeon architecture). For *titanic*, the best actual speedup is 5.86 for 10 workers, versus a best predicted speedup of 5.0 (also for 10 workers). The cost model also gives an almost perfect prediction, giving a lower bound between 6–12 workers. For *ladybank*, the best actual speedup is 8.03 for 10 workers, versus a predicted speedup of 5.26 for 10 workers. This is less accurate than for the AMD architecture, perhaps showing some additional complexities in the memory model that would repay further investigation. The cost model does, however, correctly predict a lower bound on the speedup in all cases.

6.3 Matrix Multiplication

The product of two matrices, A and B is defined as:

$$(AB)_{i,j} = \sum_{k=1}^{m} A_{ik}.B_{kj}$$

As for Cholesky decomposition, this can also be implemented using a task farm. with Fig. 7 giving speedup results for 1024×1024 and 2048×2048 matrices on both *titanic* and *ladybank*. For *ladybank*, the 1024×1024 example gives a speedup of 7.55 on 10 cores, versus a prediction of 6.33. Once again, the cost model correctly predicts lower bounds on speedup, predicting, for example, a speedup of 5.98 on 16 cores versus an actual speedup of 6.12. For the 2048×2048 execution, the best speedup is 3.54 on 14 cores, versus a predicted speedup of 3.04. For *titanic*, the 1024×1024 example gives a best speedup of 16.7 on 24 cores versus a predicted speedup of 15.65. Likewise, for the 2048×2048 example, we obtain a best speedup of 14.19 versus a predicted speedup of 11.46. In both cases, overall speedups are better for smaller matrices. This is due to lower communication costs. As with the Cholesky example, in all cases our cost models correctly predict a lower bound on speedup. In the best case, the prediction is within 3 % of the actual speedup.

6.4 QuickSort

Figure 7 shows speedup results for a *divide-and-conquer* implementation of the classical QuickSort algorithm mapped to a task farm. All instances sort the same 10^9 element list of randomly generated integers, with a threshold size of 10^5. The best actual speedup on *ladybank* was 12.72 versus a predicted speedup of 9.29.

For 8 and 10 workers, we observe a super-linear speedup of 10.87 for 8 workers and 11.2 speedup for 10 workers. While we do not have a precise explanation for this, our experiments showed that the effect is repeatable, and is presumably therefore some hardware effect. What is important is that our cost model correctly predicts a lower bound on speedup, showing a similar curve to the actual executions, with a knee at 8 workers. For *titanic*, speedup increases well from 1 to 10 workers, with an actual speedup of 6.6 on 10 workers versus a predicted speedup of 6.8 – one of the few cases where the cost model yields a (slight) over-estimate of the speedup. Beyond this point, speedup still improves, but only slightly, up to a maximum of 7.4 speedup on 24 cores (versus a prediction of 6.4). As with the other examples, the cost model closely predicts the actual speedup.

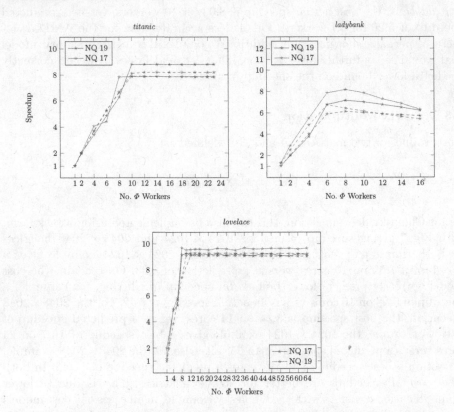

Fig. 8. Speedups for NQueens $(g \mid \Phi(s))$ on *ladybank* and *titanic*. Dashed lines are predictions.

6.5   NQueens

NQueens involves placing n queens on an $n * n$ chessboard, so that no two queens may attack each other, according to the usual rules of chess. The solution requires that no two queens occupy the same row, column or diagonal. In our implementation, we have modelled the NQueens problem as a two-stage pipeline, $g \mid \Phi(s)$, where g is the stage that generates all positions of the queens, and s is the stage that solves the position. Speedup results for this implementation of NQueens are shown in Fig. 8 for a 15×15 board, which produces 15 possible positions in the first stage, and for a 19×19 board, which produces 17 possible queens in the first stage. As with the other examples, the predicted speedups closely model the actual speedups. There is one over-prediction: for 6 workers on the 19×19 board, where we observe a discontinuity in the actual speedup. Otherwise, the predictions closely mimic the actual results that we obtain. The best speedups are 8.8 for 10 workers on a 19×19 board and 7.87 for 10 workers on a 17×17 board.

6.6   Comparison with Other Techniques

In order to demonstrate the efficiency of our queue and locking implementations, we have compared speedups for the Image Convolution example against those for two other state-of-the-art parallel implementations: OpenMP [9] and Fast-Flow [10]. Figure 9 compares all three implementations using 2048×2048 images on *titanic*. The top graph compares our implementation against FastFlow, and the bottom against OpenMP. In both cases, the speedup from our implementation is shown using solid lines, and the dashed lines show the speedups for FastFlow/OpenMP. All speedups were measured against the same sequential implementation (which took 768.75 s). In order to correctly compare the parallelism structures, we implemented the same two-stage pipeline in all three systems, using equivalent *farm* and *pipeline* structures in FastFlow, and two dynamic **for**-loops separated by a barrier synchronisation in OpenMP. The barrier is necessary to avoid the second stage processing images that have not yet been generated by the first stage, and is a natural translation of the parallelism structure that is used in the other implementations. As Fig. 9 shows, the Fast-Flow implementation is comparable to ours up to 8 workers for Φ_2 and for all versions of Φ_1, but speedup reduces drastically after this point, whereas our implementation scales well up to 23 cores. The FastFlow implementation also carries further overhead in the form of two additional dedicated cores per task farm, which is not needed in our implementation. The OpenMP implementation scales better than FastFlow, but begins to flatten out at about 6 Φ_2 workers, where our implementation continues to scale well to 23 cores.

7   Related Work

Timing issues are critically important for parallel and concurrent execution, and they have therefore been widely studied in the literature. However, despite their prevalence in real hardware, very little work considers relaxed memory models. This paper represents the first attempt of which we are aware to consider the precise impact of relaxed memory models on functional correctness, deadlock and timing. Correctness of the spin-lock protocol on x86-TSO has been previously

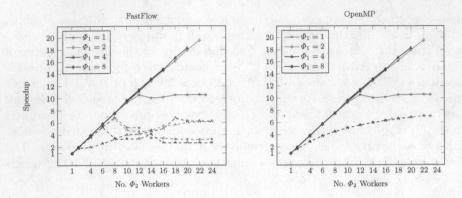

Fig. 9. Speedups for our implementation (solid lines) versus FastFlow (above, dashed lines), and OpenMP (below, dashed lines); Convolution of 500 images, size 2048×2048 (*titanic*). The baseline sequential performance is identical in each case.

proved using a semantic criterion (*triangular-race freedom* [11]) derived from the *x86*-TSO model used here [5,6]. In contrast, our new proof proves functional correctness and deadlock freedom directly over the operational model of the actual *x86-64* instructions [5,6]. Our proof has been composed with a proof for queues and thence skeletons, and furthermore clearly isolates necessary hardware assumptions for progress. We also consider, for the first time, the crucial issue of execution time. Burckhardt *et al.* [12] verify spin-locks on TSO by adapting a modified version of linearisability, and thus implicitly obtain compositionality. However, they ignore the progress and fairness constraints that we uncover, and also do not treat execution time.

Specific algorithmic skeletons are frequently associated with timing cost models [13]. However, these are obtained through measurement rather than being systematically derived from machine-level models, as here. Much of the work on developing cost models for parallel execution has focused on data parallelism. The Parallel Random-Access Machine (PRAM) execution model [14] acts as a theory of complexity for parallel algorithms on idealised shared-memory SIMD machines. In the basic PRAM execution model, basic computations and shared-memory accesses are both assumed to take unit time. Unfortunately, PRAM costs underestimate actual machine execution costs, but in an unpredictable way [15]. The Bulk Synchronous Parallel (BSP) model [16] extends the PRAM model in a more realistic way, introducing a synchronising communication step after each set of computation steps. Lisper [17] has investigated the use of a Bulk Synchronous Parallel skeleton for determining worst-case execution times, but only informally, and not in the context of real processor models. Skillicorn and Cai have likewise developed a high-level cost calculus for data parallel computations [15], based on the *shape* of data structures, and known properties of primitive parallel operations, but have not based this on a strong machine-level semantics. Blelloch and Greiner have demonstrated provable time and space bounds for nested data parallel computations in NESL [18]. To date, none of these models have therefore been derived from first principles for real multi-core architectures with relaxed-memory models. The model we give here thus represents a significant step in determining accurate cost models for algorithmic skeletons and other structured parallel forms on modern processor architectures.

The C/C++11/C++14 standards provide atomic operations that support various kinds of weak memory models [19, 20]. They could be used to implement the *Fence* and *Exchange* operations, but do not support higher-level parallel structures, such as the structured algorithmic skeletons that we have used here. Boehm and Adve [21] consider the foundations of this concurrency model. Adve and Boehm [22] give a useful survey of weak memory models as of 2010.

8 Conclusions

This paper has developed new proofs of functional correctness for *x86-64* multicores and used these to derive cost models for structured parallel programs. For the first time, we have direct operational proofs of functional correctness and deadlock freedom for standard locking and queuing algorithms under the relaxed memory model used by common *x86-64* multicores. We have used these to build accurate cost models for parallel programs that are structured using common algorithmic skeletons. Our predictions match very closely to actual results: in most cases predicting a lower-bound to within 7 % accuracy.

8.1 Limitations/Further Work

There are a number of obvious limitations to the work described here that would repay further work. Firstly, the queues that we have used here are unbounded. Extending our work to consider bounded queues should not be technically difficult, but would add some complexity to the queue definitions and would also require us to consider how "back-pressure" [23] from the demand on one queue impacts the production of values that feed that queue. Secondly, although it is very commonly used in practice, the locking mechanism that we have used here carries some, possibly significant, cost: by definition, when a processor is executing a spin-lock, it is wasting energy and not doing any useful work. More efficient notification or "lock-free"[2] techniques would obviate this at the cost of a significantly more complex proof and rather more complex cost model. We have therefore chosen not to do this here. Thirdly, and more seriously, as with most work on algorithmic skeletons, we have not considered any form of feedback, as in FastFlow's "farm/pipeline-with-feedback" skeletons [10]. Incorporating skeletons with feedback would allow the construction of more complex parallel programs, but would require us to determine fixed-points in the skeletons and to solve the resulting timing recurrence relations. Fourthly, we have ignored some hardware

---

[2] In the sense that locks are not visible to the programmer, rather than they are not used by the hardware.

effects. In particular, while we have accounted for the behaviour of the cache and associated hardware when determining functional correctness and deadlock properties, we have not attempted to precisely model memory access behaviour in determining execution times, but have assumed that the sequential cost model properly accounts for such costs. We have also not modelled the processor instruction pipeline. Neither issue will affect functional correctness, but could, obviously, impact low-level timing accuracy.

Acknowledgements. This work has been partially supported by the EU Horizon 2020 grant "RePhrase: Refactoring Parallel Heterogeneous Resource-Aware Applications – a Software Engineering Approach" (ICT-644235), by COST Action IC1202 (TACLe), supported by COST (European Cooperation in Science and Technology), and by EPSRC grant EP/M027317/1 "C^3: Scalable & Verified Shared Memory via Consistency-directed Cache Coherence".

References

1. Cole, M.I.: Algorithmic Skeletons: Structured Management of Parallel Computation. MIT Press, Cambridge (1991)
2. González-Vélez, H., Leyton, M.: A survey of algorithmic skeleton frameworks: high-level structured parallel programming enablers. Softw. Pract. Experience **40**(12) 1135–1160 (2010)
3. Danelutto, M., Torquati, M.: A RISC building block set for structured parallel programming. In: Proceedings of the 21st Euromicro International Conference on Parallel, Distributed, and Network-Based Processing (PDP 2013), pp. 46–50 (2013)
4. Lamport, L.: How to make a multiprocessor computer that correctly executes multiprocess programs. IEEE Trans. Comput. **C-28**(9), 690–691 (1979)
5. Sewell, P., Sarkar, S., Owens, S., Zappa Nardelli, F., Myreen, M.O.: x86-TSO: a rigorous and usable programmer's model for x86 multiprocessors. CACM **53**(7), 89–97 (2010)
6. Owens, S., Sarkar, S., Sewell, P.: A better x86 memory model: x86-TSO. In: Berghofer, S., Nipkow, T., Urban, C., Wenzel, M. (eds.) TPHOLs 2009. LNCS, vol. 5674, pp. 391–407. Springer, Heidelberg (2009). doi:10.1007/978-3-642-03359-9_27
7. Intel: Intel 64 and IA-32 Architectures Software Developer's Manual vol. 3A: System Programming Guide, Part 1, §8.2.2. Intel (2013)
8. Papamarcos, M.S., Patel, J.H.: A low-overhead coherence solution for multiprocessors with private cache memories. In: Proceedings of the ISCA 1984: 11th Annual International Symposium on Computer Architecture, pp. 348–354. ACM (1984)
9. Chapman, B., Jost, G., van der Pas, R.: Using OpenMP: Portable Shared Memory Parallel Programming (Scientific and Engineering Computation). The MIT Press, Cambridge (2007)
10. Aldinucci, M., Danelutto, M., Kilpatrick, P., Torquati, M.: Fastflow: high-level and efficient streaming on multi-core. In: Programming Multi-core and Many-core Computing Systems. Parallel and Distributed Computing (2012)
11. Owens, S.: Reasoning about the implementation of concurrency abstractions on x86-TSO. In: D'Hondt, T. (ed.) ECOOP 2010. LNCS, vol. 6183, pp. 478–503. Springer, Heidelberg (2010). doi:10.1007/978-3-642-14107-2_23

12. Burckhardt, S., Gotsman, A., Musuvathi, M., Yang, H.: Concurrent library correctness on the TSO memory model. In: Seidl, H. (ed.) ESOP 2012. LNCS, vol. 7211, pp. 87–107. Springer, Heidelberg (2012). doi:10.1007/978-3-642-28869-2_5

13. Hamdan, M.M.: A Survey of Cost Models for Algorithmic Skeletons. Technical report, Heriot-Watt University (1999)

14. Fortune, S., Wyllie, J.: Parallelism in random access machines. In: Proceedings of the STOC 1978: 10th Annual ACM Symposium on Theory of Computing, pp. 114–118. ACM (1978)

15. Skillicorn, D.B.: A cost calculus for parallel functional programming. J. Parallel Distrib. Comput. **28**, 65–83 (1995)

16. Valiant, L.G.: A bridging model for parallel computation. Commun. ACM (CACM) **33**(8), 103–111 (1990)

17. Lisper, B.: Towards parallel programming models for predictability. In: Proceedings of the WCET 2012: 12th International Workshop on Worst-Case Execution Time Analysis. OpenAccess Series in Informatics (OASIcs), vol. 23, pp. 48–58 (2012)

18. Blelloch, G.E., Greiner, J.: A provable time and space efficient implementation of NESL. In: Proceedings of the ICFP 1996: ACM SIGPLAN International Conference on Functional Programming, pp. 213–225 (1996)

19. Becker, P. (ed.): Programming Languages – C++. ISO/IEC (2011)

20. Williams, A.: C++ Concurrency in Action: Practical Multithreading. Manning Publications (2012)

21. Boehm, H.J., Adve, S.V.: Foundations of the C++ concurrency memory model. In: Proceedings of the PLDI 2008: 29th ACM SIGPLAN Conference on Programming Language Design and Implementation, pp. 68–78 (2008)

22. Adve, S.V., Boehm, H.J.: Memory models: a case for rethinking parallel languages and hardware. Commun. ACM (CACM) **53**(8), 90–101 (2010)

23. Collins, R.L., Carloni, L.P.: Flexible filters: load balancing through backpressure for stream programs. In: Proceedings of the EMSOFT 2009: ACM SIGBED International Conference on Embedded Software, pp. 205–214 (2009)

Author Index

Printed in the United States
By Bookmasters